Building Failures

DIAGNOSIS AND AVOIDANCE

Building Failures

DIAGNOSIS AND AVOIDANCE

W. H. RANSOM

London New York
E. & F. N. SPON

First published in 1981 by E. & F. N. Spon Ltd
11 New Fetter Lane, London EC4P 4EE

Published in the USA by E. & F. N. Spon
29 West 35th Street New York, NY 10001

Second edition 1987

© 1987 W. H. Ransom

Printed in Great Britain at the University Press, Cambridge

ISBN   0 419 14260 6 (HB)
        0 419 14270 3 (PB)

British Library Cataloguing in Publication Data

Ransom, W. H.
    Building failures: diagnosis and
    avoidance.——2nd ed.
    1. Building failures——Great Britain
    2. Buildings——Great Britain——Protection
    I. Title
    690'.24   TH441
    ISBN 0-419-14260-6
    ISBN 0-419-14270-3 Pbk

Library of Congress Cataloging in Publication Data

Ransom, W. H.
    Building failures.
    Bibliography: p.
    Includes index.
    1. Building failures.   I. Title.
TH441.R36   1987   690   87-12418
ISBN 0-419-14260-6
ISBN 0-419-14270-3 (pbk.)

Contents

Acknowledgements

Improvements in building techniques, understanding of the reasons for, and the prevention of, failures, the encouragement of sensible innovation and the development of safe standards owe much to the Property Services Agency, the Building Research Establishment, the British Board of Agrément and the British Standards Institution. So, too, does this book. I wish also to make specific acknowledgement to the Property Services Agency for permission to use Figs 4.2, 8.2, 10.1, 10.5, 10.6 and 11.1, to the Director of Building Research for Figs 2.1. 3.1, 3.2, 3.3, 3.4, 4.1, 4.4, 6.1, 6.4, 6.5, 6.6, 7.5, 7.6, 7.7, 7.8, 7.9, 7.10, 9.1, 9.2 and 11.2. Figure 8.1 is reproduced by kind permission of the Brighton Evening News and Argos. Figures 6.2 and 7.3 are based on figures in publications by the Building Research Establishment. I wish also to thank the Chartered Institution of Building Services Engineers for permission to include Table 2.1, to the British Standards Institution for agreeing to the publication of Tables 4.1 and 4.2 and to Mitchell Beazley Publishers for the use of information resulting in Table 5.2.

W.H.R.

1
Introduction

Most building defects are avoidable: they occur, in general, not through a lack of basic knowledge but by non-application or mis-application of it. Knowledge seems to become mislaid from time to time. Those with long memories, and those whose business it is to make a particular study of building defects, are often struck by the re-emergence of problems which have been well researched and documented. Certain basic properties of materials, such as their ability to move through changes in temperature and moisture, seem to be overlooked and a rash of difficulties occurs. A call goes out for more research but, in truth, all that is usually needed is a good system for the retrieval of information, a better procedure for its dissemination and, most important, the realization that an information search is desirable.

Current training in design tends to concentrate on what to do rather than what not to do. A similar situation exists in training in constructional techniques, where the craftsman is instructed how best to undertake a particular operation but, to a lesser extent, in the dangers of deviation from an accepted technique. Understanding of the likelihood of defects through inadequate design or construction is taught implicitly rather than explicitly. The level and nature of defects in building construction currently encountered suggest that more guidance is required on the avoidance of failures. A need is seen, too, for such guidance to be a positive part of a training curriculum. Indeed there are good arguments for suggesting that, as the first essential in design and construction is to ensure that the structure provided is stable and durable, specific education in the avoidance of failure should be a major part of any design and construction syllabus. The purpose of this book is to provide this positive guidance in a suitably compressed form. It does not set out to describe every possible way in which a building may become defective: such a task would scarcely be possible and certainly would not be particularly helpful. It seems better to aim at identifying the principal defects and their causes which, if wholly eliminated, would prevent the great majority of the defects which currently occur, save occupants of buildings much annoyance and discomfort, and reduce the national bill on maintenance and repair by scores and, possibly, by hundreds, of millions of pounds annually.

The book aims to identify the nature and cause of important defects occurring in buildings, with emphasis on those affecting the fabric of a building and its associated services. It does not deal with issues of aesthetics, lighting, or thermal or acoustical comfort. While concerned primarily with the avoidance of defects, the text also gives guidance to aid in their correct diagnosis when, unfortunately, the situation demands cure rather than prevention. Except in a general way, the repair of such defects is not covered. Any one specific failure needs a detailed examination to decide on the most appropriate repair, for this depends not only upon technical considerations but also upon the type of building and its age, and upon related economic and social considerations. There are few standard solutions to problems of repair.

Most defects occur through the effects of external agencies on building materials and the two succeeding chapters consider in some detail the nature of these and their effects on the materials commonly used in building. These agencies include the principal components of the weather, namely, solar radiation, moisture and air and its solid and gaseous contaminants; biological agencies, in particular fungi and insects; ground salts and waters; and manufactured products used in conjunction with building materials, for example, calcium chloride. Moisture occupies a central role, as the villain. Work initiated by the Joint Working Party on Heating and Energy Conservation in Local Authority Housing has shown that there may be as many as two and a half million dwellings in the UK suffering from dampness and in two-thirds of these the dampness is caused by condensation. The main sources of moisture and ways in which the amounts present may be minimized are dealt with in Chapter 4. Special emphasis is given to the cause and effects of condensation, and how the risks may be avoided or reduced. Condensation, particularly in local-authority dwellings, can truly be said to have been the greatest single cause of human discomfort in dwellings since the end of the Second World War. The elements of building structure are then dealt with, starting with foundations and progressing logically upwards to roofs and parapets, passing on the way, floors, walls, cladding and external joinery. The avoidance of defects in building services has a chapter to itself. The book concludes with a more speculative chapter dealing with failure patterns and control. This attempts to relate defects to problems associated with the structure of the industry, to the dissemination of information and to particular difficulties which result from rapid innovation. Current control methods are outlined and a possible strategy is suggested for improving control, quality and reliability.

The intention and hope is that this book will provide positive guidance to the student designer and builder on how to avoid the principal building defects. It includes no complex scientific concepts and requires no special knowledge of science. Though concerned more with normal building than with major civil engineering construction, much of the text is of relevance to structural engineers also, particularly those parts dealing with the properties of the structural materials, with foundations and with cladding. The point was made at the beginning that knowledge gets mislaid: a further aim of this book is to serve as an aide-mémoire for practising designers and builders. For this reason, it has been kept concise, and is illustrated to give visual emphasis to some of the more important defects which can occur. These illustrations and parts of the text which describe the likely appearance of failures may assist surveyors and maintenance personnel, too, by steering them towards the probable cause of a failure. Though the essential aim is to avoid failure, once it has occurred and maintenance is needed, it is hoped the book will help both in identifying the cause and in preventing the adoption of the wrong remedial action. It may also help the maintenance engineer and surveyor by putting the severity of a failure and its consequences into a reasonable perspective and so prevent over-reaction to the event, which is not uncommon, particularly with foundation problems. If the book succeeds only partly in these ambitions it will, nevertheless, save both money and misery.

2

Agencies causing deterioration

When building materials and components are transported, stored on site and used in a structure, they are subjected to the effects of a number of agencies, some of which may influence adversely their durability and performance and, thereby, have a major bearing on the possibility of their premature failure. The action of the weather, or the external climate, is foremost amongst these. It is often taken to affect only those materials exposed externally. However, the distinction between the external and internal environments in a building is not always clear-cut. There are partially protected areas still open to the weather to some extent. Outside air enters a room through an open window and sunlight is filtered through window glass. Weather changes in protected or internal environments are usually the same in type but slower in action than those taking place outside. Buildings themselves cause modifications to the weather and to the micro-climate, considerable differences in which may occur in quite short distances. The micro-climate is, indeed, of particular significance and, because of the large number of individual circumstances which can cause its modification, is still an area of much uncertainty and a fruitful one for detailed investigation.

The principal components of the weather include:

(1) Radiation – from the sun and from the rest of the sky; varying in amount, frequency, direction, intensity and spectral composition.
(2) Rain – varying in direction, droplet size, quantity, intensity, duration, temperature and distribution.
(3) Solidified water – snow or hail; varying in frequency, direction, shape, size, amount, terminal velocity.
(4) Air and its gaseous constituents – in particular, water vapour, oxides of sulphur, oxygen and carbon dioxide.
(5) Solid and liquid contaminants of air – dirt, tar and oil particles, salts;

varying in composition, distribution, ease of attachment and detachment, chemical effects. (Salt spray near the coast is an important example.)

Durability and performance are also affected by biological agencies (of which moulds, fungi, bacteria and insects are the most important), ground waters and salts, and manufactured products, for example, calcium chloride. It is worth making the point that durability is not an inherent property of a material: different materials have different effective durabilities due in part to each of the following – their physical and chemical properties, the function each has to perform and their position on, or in, a building. In practice, each building material or combination of materials tends to respond differently to the influences outlined above, many of which may be active at any one time. These influences are often inter-related in a complex way and may reinforce or oppose one another in different materials. Thus, coincident strong sunlight and dew have a particularly damaging effect upon paint films. On the other hand, unobstructed sunlight following dew deposited on metal will assist evaporation and so reduce the likelihood of corrosion. One agency alone may exert very different effects, depending on its form or intensity. For example, water in the form of rain washing over a surface can retard or prevent mould growth, but moisture in the form of repeated condensation can be highly conducive to its formation. For any particular situation, it is necessary to assess the likely combination of agencies and their effects upon durability and performance, and succeeding chapters seek to make this assessment for the most common building situations. As a basis to considering these more complex inter-relationships, it is, nevertheless, useful to consider separately the agencies mentioned and their general effects.

2.1 SOLAR RADIATION

Solar radiation is received at the Earth's surface both directly and as long-wave diffuse sky radiation. The proportion of diffuse sky radiation to total radiation received is considerable and cannot be neglected: indeed, it can exceed direct radiation. Solar radiation is absorbed when it strikes an opaque surface. Most building materials are opaque and their absorptivity (the ratio of the radiation absorbed to the incident radiation received) varies, depending upon the nature and colour of the surface. Black non-metallic surfaces have high absorption. Some values for building materials are shown in Table 2.1. The absorption by some

Table 2.1 Absorptivity to solar radiation of some common materials

Material	Absorption coefficient (clean materials)
Aluminium	0.2
White sand–lime bricks	0.4–0.5
Limestone	0.3–0.5
Concrete tiles	0.65
Dark fletton bricks	0.65
Mortar screed	0.8
Asphalt	0.9

This table was originally published in Section A6 of the CIBSE Guide by permission of the Chartered Institution of Building Services Engineers.

materials of bands of short-wave solar radiation (referred to generally as the ultra-violet) can lead to degradation. Such degradation is confined to organic materials, in particular, to plastics, some paints and bituminous-based materials.

2.1.1 Temperature effects

The absorption by surfaces of solar radiation is accompanied by a rise in temperature. Building surfaces can also emit long-wavelength radiation and, in so doing, cool. The drop in temperature can be considerable, particularly on clear nights when radiation from black surfaces, such as asphalt roofs, can cause surface temperatures to fall well below shade air temperature. A rise in temperature leads to an increase in the rate of reactions and can accelerate many degradation processes. (An increase of 10°C doubles the rate of many chemical reactions.) High temperatures, in themselves, also lead to high rates of evaporation and volatilization. Loss of volatiles from bituminous compositions, some plastics, mastics and sealing compounds can cause shrinkage and brittleness. Evaporation of water from cement mixes can lead to early weakness, poor adhesion and cracking. Phase changes may occur, the best-known example being that which occurs in high-alumina cement, the change being closely associated with a loss in strength. Some building materials, for example bitumen, soften or melt with high temperatures. In contrast, temperatures that are permanently below freezing can be highly favourable, which is often not recognized, for they ensure the absence of all cyclic thaw–freeze effects, of leaching by rain and of liquid moisture migration. Many degradation processes, too, are slowed down.

Temperature changes cause dimensional changes in materials, particularly when the coefficient of expansion is high as, for example, with aluminium and some plastics. These changes cause stresses which, if not accommodated, can exceed the strength of some materials and cause distortion or rupture. Temperature changes can be quite sudden. Sunlight breaking through a frost-laden fog can heat up a surface very rapidly. Rain falling on a sun-heated surface applies a severe quenching shock. Brittle coatings and joints between dissimilar materials can then undergo the first initial breakdown, leading to subsequent deterioration. In the United Kingdom, air temperatures can change by over 20°C between night and day, and by over 50°C between maximum summer temperatures and minimum winter ones. The maximum temperature changes on the surface of building materials and the rates of temperature change are often even greater. Black surfaces which are good

Table 2.2 Thermal expansion of some common building materials

Material	Approximate coefficient of linear expansion per °C ($\times 10^{-6}$)	Unrestrained movement for 50°C change (mm/m)
Bricks and tiles (fired clay)	5–6	0.25–0.3
Limestone	6–9	0.3–0.45
Glass	7–8	0.35–0.40
Marble	8	0.40
Slates	8	0.40
Granite	8–10	0.40–0.50
Asbestos cement	9–12	0.45–0.60
Concrete and mortars	9–13	0.45–0.65
Mild steel	11	0.55
Bricks (sand–lime)	13–15	0.65–0.75
Stainless steel (austenitic)	17	0.85
Copper	17	0.85
GRP	20	1.0
Aluminium	24	1.2
Lead	29	1.45
Zinc (pure)	31	1.55
PVC (rigid)	50	2.50
PVC (plasticized)	70	3.50
Polycarbonate	70	3.50

absorbers of solar radiation and powerful emitters of lower-temperature radiation can show the maximum changes in temperature and, particularly so, if insulated behind the black surface. The range of temperature changes for such surfaces may be double that of the air-temperature changes mentioned above. The coefficients of thermal expansion of some typical building materials are shown in Table 2.2, together with an indication of the unrestrained movements consequent upon a change in temperature of 50°C – an annual change which can be expected as a minimum for many materials.

2.2 MOISTURE

Moisture in solid, liquid or vapour form can be regarded as the principal agent causing deterioration. It is always present in the atmosphere and, when surface temperatures of materials fall sufficiently, condensation can occur, which may be heavy and prolonged. Even under cover, surfaces can become thoroughly wetted, metals may corrode and be brought into aqueous contact with other metals or materials which may lead to electrolytic attack, and glass may be etched. Conditions are particularly conducive to deterioration when moisture condenses in relatively inaccessible crevices from which subsequent evaporation is slow. Rain, particularly when blown by strong winds, can erode soft materials and, washing over a surface, may remove part of it in solution. Water has a high heat of vaporization and is, therefore, slow to evaporate. High precipitation, consequently, implies not only a more complete, but a more prolonged, wetting of materials. When water freezes in the pores of materials, such as brick, stone and concrete, stresses are produced which may cause spalling of the surface, general cracking or disintegration. Water frozen in the form of hail can cause pitting of some surfaces and, as snow, has to be allowed for in structural design.

Changes in relative humidity can lead to dimensional change in materials, with deformation, crazing or cracking. Prolonged low humidities can cause the dehydration of gypsum products (though such humidities do not occur naturally in the UK); prolonged high humidities aid fungal growth and the subsequent decay of organic materials. Moisture also stimulates biological activity and acts as a medium or catalyst through which reactions occur which could not otherwise take place. Because of its major role in causing or assisting failures in materials, components and structures, moisture is dealt with separately and at greater length in Chapter 4.

2.3 BIOLOGICAL AGENCIES

Attacks by fungi and insects are principally upon timber, though other materials, generally organic, can be affected. In recent years, a good deal of timber decay has been caused by the several varieties of fungi which are covered by the term 'wet rot'. Wet-rot fungi require markedly damp conditions to germinate and a continuing source of moisture for their existence. Sapwood at a moisture content of around 30% and a temperature around 20°C provides an ideal abode. Timber with a moisture content not greater than 20% is not endangered. Once the source of moisture is removed, the fungi will die and do not have the ability to spread to dry timber, or to penetrate plaster and brickwork, as does the fungus *Serpula lacrymans* (dry rot). Spores of the dry-rot fungus are generally present in the air and, given the right conditions, will germinate. Though they are too small to be seen individually by the unaided eye they can be spread by moving air in a building to settle as a rust-coloured dust. The conditions favoured are dark stagnant ones with timber of moisture content above 20% and temperatures around 20°C. Poorly ventilated sub-floor areas and situations where timber is in prolonged contact with damp materials are those where the danger of attack is greatest. The fungus grows most readily on unsaturated wood which has a moisture content of 30% or more. Once germination has occurred filaments called hyphae spread over the surface to form whitish fluffy growths or sheets known as the mycelium. The hyphae can penetrate cracks in materials such as plaster, brick and block, which in themselves do not provide nourishment, in search of further wood. The growth of the fungus over these inorganic materials is made possible by strands, much thicker than hyphae, which are formed behind the latter and convey food and moisture from the damp wood where the attack began to these hyphae. If further wood is then reached it can be attacked even if not damp for moisture is conveyed to the cellulose in the drier wood by the strands, though there is some evidence that the fungus is not efficient at conducting moisture from a wet to a dry location [1]. The affected wood loses its cellulose, its strength and its weight and cracks and shrinks often into cubic pieces.

Dark, damp and stagnant conditions are also favoured by the various moulds which can, under those circumstances, damage decorations, furnishings and fabrics, and have been the cause of many complaints in local-authority housing subject to acute condensation problems.

The common furniture beetle (*Anobium punctatum*) is responsible for most of the damage caused to timber by insect attack in the UK. The

Fig. 2.1 Rot in roof timbers.

eggs are laid in cracks and joints and when the larvae hatch they bore into the timber. They remain in the timber for around three years, growing all the while, until they reach up to 6 mm long. During this period, the insects tunnel to and fro within the timber, thus weakening it. When fully grown, they emerge from the wood as beetles, leaving circular holes approximately 1.5 mm in diameter. Plywood made from birch and bonded with natural glues is particularly susceptible to attack. More serious damage, though fortunately less wide-spread, can be caused by the house longhorn beetle (*Hylotrupes bajulus*). The beetle is generally around 10 to 20 mm long and can cause serious damage to the sapwood of softwoods, particularly to roof timbers. Hardwoods are not attacked. Unfortunately, the larval stage, during which the grub itself can grow to over 30 mm, can last up to ten years, though five to six years is more common in the UK. During that time, extensive tunnelling and damage can occur. The exit holes are oval, not circular as with *Anobium punctatum*, widely spaced and up to 10 mm wide. Timber which shows external signs of attack may well be very seriously weakened and so need total removal and destruction. Fortunately, attack by house long-horn beetle in the UK seems to be confined, so far, mainly to areas in Berkshire, Hampshire and Surrey, and the 'Approved Document' to support Regulation 7 of the Building Regulations 1985 [2] requires the special treatment of softwood against the beetle in specified areas within those counties. Damage to hardwoods, usually oak, may also be caused by deathwatch beetle (*Xestobium rufovillosum*). The beetle, however, attacks only those woods which have already been weakened by fungal attack. It is a problem generally confined to old, large structures.

2.4 GASEOUS CONSTITUENTS AND POLLUTANTS OF AIR

Sulphur dioxide is generated by the burning of fuel and concentrations in the atmosphere are greatest in large industrial areas. Great improvements have been made in recent years in preventing harmful emission of sulphur gases – concentrations in London are now around one-third of what they were just after the Second World War. Sulphur gases dissolved in rainwater, nevertheless, still rank as the most aggressive gaseous pollutant and can assist the corrosion of some metals and cause some stone to blister and to spall.

Carbon dioxide is a normal atmospheric gas of little general importance but, disssolved in rainwater, it forms a weak acid capable of slowly eroding limestone and weakening calcareous sandstones by solution of the bonding calcite. It will slowly carbonate lime formed during the hydration of cement. A change in physical properties may result. Thus, asbestos cement becomes more brittle and a lime bloom may appear on concrete products generally. The extent of carbonation in concrete can have a marked influence on the corrosion rate of reinforcement and this is considered in more detail later. As far as is known, nitrogen and the inert gases in the air do not affect building materials. The oxygen content does not vary sufficiently to cause a difference in its oxidative effect on those building materials (such as the organic ones) which are slowly degraded by it. Ozone is rather more variable in distribution. Though present only in traces, it plays the dominant part in the degradation of rubber, particularly when it is stressed, and can be presumed also to influence the degradation of mastics, bituminous compositions, paints and plastics. However, its effects are not of major importance.

2.4.1 Solid contaminants

Dirt and general grime in the atmosphere consist of inorganic dust particles, together with unburnt particles of fuel, bound together by a form of oil or other organic matter derived from fuel, including that from road vehicles. The dirt also contains some soluble salts. Such dirt is deposited on buildings and causes an adverse effect on appearance.

There is some evidence, however, that more harmful effects may result, such as an increase in the corrosion rate of metal surfaces and the deterioration of some stone surfaces.

Near the coast, the concentration of salt derived from sea-spray is high and the corrosive effects on metals can be severe. Salt particles are deliquescent and absorb moisture from the atmosphere, forming strong solutions, and they are liable to cause severe corrosion if they lodge in crevices.

In the close vicinity of particular industries, a variety of gaseous and solid contaminants may affect buildings but, in a national context, the effects are relatively unimportant.

2.5 GROUND SALTS AND WATERS

Salts, present in the ground, can rise in solution and by capillarity into porous materials with which they are in contact. Upon subsequent evaporation of the solvent (water), salts may be deposited within the pores or upon the surface of the material, the exact site of deposition depending upon the pore structure and the rate of drying. Usually, little more than efflorescence occurs, often ephemeral and of little consequence. More seriously, if magnesium sulphate is present, disintegration of renderings and masonry surfaces can occur, though failures appear to be infrequent. Peaty moorland ground waters are acidic and can cause concrete in contact with them to lose strength and to erode, the effect increasing with prolonged contact.

2.6 MANUFACTURED PRODUCTS

Manufactured materials may be used as additions to building materials, or as treatments for them, and may have an adverse effect on durability and performance, if not used with care and understanding. Calcium chloride is such a material and is commonly used as an accelerator of the hydration and development of the early strength of cement-based products, for example, mortar, concrete and wood wool. Its use has commercial advantages, in that it facilitates the early demoulding of precast elements and their removal from the factory production floor to storage and, also, the early striking of formwork from *in situ* concrete. However, it has a corrosive effect on metals and severe damage has been caused to reinforced concretes and to prestressed concrete through its use. This problem is considered in more detail in Chapter 3.

Inorganic salts may be used as fire retardants and as preservatives in wood. While these have no adverse effects on the wood itself, they can, under some environmental conditions, assist the corrosion of metal fasteners used in timber construction.

2.7 JUXTAPOSITION OF MATERIALS AND COMPONENTS

The use of some building materials in proximity to each other can lead to weathering effects, with consequent maintenance expenditure. Staining is probably the most familiar effect and can be caused by design details, resulting in dirt deposition in some places and its removal in others: unsightly streaks on the building are often the result. Products of metal corrosion can leave rust stains on concrete. Efflorescence from brickwork gives a particularly common stain.

Transmission of a soluble component from one part of a building to another, however, may cause not only staining but more serious trouble. Thus sulphates from bricks can cause breakdown of mortar, corrosion of metals or disintegration of stone. Most timbers are acidic, particularly the heartwood of oak and of Douglas fir, due to the presence of volatile acetic acid, and can corrode some metals. Western red cedar shingles, through the presence in the wood of water-soluble organic derivatives, and lichens on roofs can both attack metal rainwater goods and flashings. The external use of aluminium, copper, phosphor-bronze and other non-ferrous metals, and of stainless steel, needs great care because of the often unforeseen opportunities for electro-chemical corrosion. Washings from limestone onto some sandstone plinths and mouldings can cause deterioration of the latter. It should be emphasized that different materials which, in isolation, may possess adequate durability can deteriorate markedly when brought into aqueous contact with one another. In general, such dangers have been well documented but, with the introduction of newer materials into construction, less well-known problems may arise. For example, polystyrene in contact with PVC can lead to loss of plasticizer from the latter, causing it to become brittle and the polystyrene to 'shrink' away.

3

Durability of materials

3.1 ASBESTOS CEMENT

Asbestos cement is made from asbestos fibres, ordinary Portland cement and water. The cement hydrates and sets around and between the asbestos fibres, which act as reinforcement. The fibres are based on the mineral chrysolite and have an average length of around 5 mm. The relative proportions of asbestos fibre and cement used vary with the particular product to be made: for roof sheets and slates the proportion of asbestos is around 10% by weight of the dry materials. Admixtures which may also be used in small quantities are principally colouring pigments and cellulose fibre.

The coefficient of thermal expansion is small (see Table 2.2). Changes in moisture content cause far greater movement. When dry asbestos-cement slates are wetted, they can expand by between 0.1 and 0.3% of their initial dry length: the older the slates, the smaller is their movement. Carbon dioxide, always present in the atmosphere, carbonates the lime formed by the hydration of the Portland cement and this carbonation causes shrinkage of the exposed asbestos cement. Whether the effects of moisture causing expansion, or carbonation causing shrinkage, will predominate is dependent largely upon the age of the asbestos cement and the number of wetting and drying cycles to which it has been subjected already. The greater these are, the greater will become the predominance of the effects of carbonation and, eventually, normal exposure will result in overall shrinkage.

Asbestos-cement products are highly resistant to atmospheric pollution and it is mainly physical changes which make them more prone to damage. Sheets harden with age through hydration of the cement matrix, which can continue for a long while. Atmospheric carbonation decreases impact strength which, over a long period, can fall to half its initial value. The long-term effect of natural weathering, therefore, is for sheets to become more brittle, and more liable to damage by impact.

Both old and new asbestos cement can be damaged by rough handling during transport and erection.

When the two faces of a sheet are subjected to widely differing rates of atmospheric carbonation, differential shrinkage occurs. Some instances are known where this has been sufficient to cause cracking, for example, where the external face has been painted or surface-coated, which retards carbonation, while the internal face has been left un-painted. Carbonation shrinkage of asbestos cement, together with the tendency to expansion through cement hydration and moisture effects, can cause warping, which is enhanced by any differential movement between the two surfaces. When the external face of asbestos-cement sheets or slates is treated with a surface coating, the internal face should, at least, be primed. This will help to reduce differential moisture movement and carbonation shrinkage between the two faces. Cracking and warping of sheets can occur even when they are correctly fixed to sound supports, unless these precautions are taken.

Asbestos cement is not susceptible to attack by insects but both algal and fungal growths are common on the normal unpainted product. The surfaces become slightly softened, with resultant darkening and dis-colouration, which is unpleasant aesthetically and, with roofing and cladding sheets, may be undesirable because of the increased ability to absorb solar radiation.

Because of the disabling lung disease asbestosis associated with the use of asbestos and its products and given much prominence in a television programme in 1978, manufacturers of asbestos-cement prod-ucts have moved rapidly towards the production of non-asbestos substitutes. The trend will certainly continue but asbestos-cement products are still on the market, in particular corrugated cladding and roofing sheets and roofing slates. Such sheets and slates are covered by BS 690 [3].

3.2 ASPHALT AND BITUMEN

A range of products is used in the building industry, referred to broadly as bituminous or asphaltic, and the terminology can be confusing. The following definitions are based on those contained in BS 6577 [4]:

Bitumen – a viscous liquid, or a solid, consisting essentially of hydro-carbons and their derivatives, which is soluble in carbon disulphide. It is substantially non-volatile and softens gradually when heated. It is black or brown in colour, and possesses waterproofing and adhesive proper-ties. It is obtained by refinery processes from petroleum and is also found as a natural deposit or as a component of naturally occurring asphalt, in which it is associated with mineral matter.

Asphalt – a mixture of bitumen with a substantial proportion of inert mineral matter. Asphalt may occur in nature as natural rock asphalt, which is a consolidated calcareous rock impregnated with bitumen exclusively by a natural process. It may also occur as lake asphalt (Trinidad), where it is in a condition of flow or fluidity. Mastic asphalt, which is the asphalt used in building, is a type of asphalt in which the mineral matter is suitably graded in size.

Exposure to sunlight over a substantial period causes mastic asphalt and bitumen to harden and to shrink, the shrinking often resulting in slight surface crazing. This effect, which is due to short-wave solar radiation, is assisted by atmospheric oxidation. Being black or dark grey in colour, its absorption of solar radiation is high (see Table 2.1) and the high temperatures reached as a consequence can, if the material is not of the right grade, cause softening and flow. Even when correctly graded materials are used, thermal expansion and contraction can be high, through the wide temperature changes resulting from high absorptivity and high emissivity. When such changes are slow, these movements can be accommodated without internal damage but rapid movements through sharp changes in temperature can cause cracking, particularly in cold weather. Under mechanical stress, bitumen products can flow. Asphalt and bitumen are not affected by biological agencies or by pollution but contact with oil can be damaging. Moisture has no direct adverse effect – the materials are used essentially for waterproofing – but moisture-vapour pressure can cause blistering. This problem, and others affecting the use of asphalt and bitumen for roofing, are dealt with later.

3.3 BRICKS AND TILES

Most bricks used are of burnt clay or calcium silicate, the former being classified in BS 3921 [5] and the latter in BS 187 [6]. Clay bricks suitable for general building purposes are described as 'common'. Facing bricks are specially made or selected for their attractive appearance. Engineering bricks are strong bricks of low permeability.

The coefficient of thermal expansion of clay bricks is shown in Table 2.2. This movement is greatly exceeded by an irreversible moisture movement which occurs when newly fired bricks absorb moisture. Typical expansion of an individual brick can be around 0.1% and even close to 0.2%, in some cases; around half that movement, however, will occur in the first week after manufacture. Such an expansion can cause problems, particularly when associated with opposing movements in structures, and this is considered in more detail further on, where cladding is dealt with. A good deal of potential difficulty can be avoided by not using bricks fresh from the kiln: brickwork, as opposed to individual bricks, suffers less expansion – in general, just over half the amount. A normal reversible wetting and drying movement of some 0.02% occurs in addition to this irreversible expansion.

Clay bricks, which have a water absorption of less than 7% by weight, have a high resistance to damage by frost. It should not be assumed, however, that bricks with a high water absorption have a low resistance, for resistance depends not only on total porosity but also on pore structure and, in particular, upon the proportion of fine pores present, the resistance to attack increasing as this proportion decreases. Experience has shown that frost damage in the UK is unlikely in walls between eaves and damp-proof course (DPC) level. Whether this will continue to be true for highly insulated buildings, in which the outer leaf may be colder than hitherto, remains to be seen. Below DPC level and in parapets and free-standing walls, frost damage does occur and care is needed to select bricks of appropriate frost resistance. Information on such matters is available, in particular, from the Brick Development Association.

Many burnt clay bricks contain small amounts of salts, usually sulphates, and these may crystallize at the surface to give a white deposit, referred to as efflorescence. This is noticeable, particularly, in dry periods following building and is usually harmless though, in exceptional cases, some crumbling of the surface may occur. This possibility is virtually restricted to underfired bricks, which not only present a weaker surface to attack but are also likely to contain larger amounts of deleterious salts. Some bricks may contain iron salts which can produce rust stains and these are accentuated when such bricks are left exposed to rain before building. The stains affect appearance only and have no harmful effects on general durability.

Calcium silicate bricks are made not by burning clay but by reacting lime under steam pressure with a suitable source of silica, usually sand or crushed flint. In BS 187 [6], calcium silicate bricks are classified partly by their drying shrinkage, which typically lies between 0.025 and 0.035%. The coefficient of thermal expansion is around 14×10^{-6} per °C. Their movement is greater than for clay bricks, which necessitates a greater allowance for movement in the design of brickwork. Unlike burnt clay bricks, there is a marked correlation between frost resistance

Fig. 3.1 Efflorescence on brickwork.

and strength. The stronger classes shown in BS 187 are required for the more exposed situations.

Calcium silicate bricks are free from the salts which cause efflorescence in clay bricks. Exposure to atmospheric carbon dioxide causes slight hardening and shrinkage but is of little practical significance. Sulphur dioxide causes decomposition of the calcium silicate bonding material and this can cause general weakening and blistering in a very polluted environment. Sea spray is detrimental, causing erosion of the surface and also reducing resistance to frost attack.

Concrete bricks conforming to BS 6073 (Part 1) [7] have drying shrinkage which may range from 0.04 to 0.06%, with wetting expansion not greater than 0.06 to 0.08%.

Roofing tiles are severely exposed. Their resistance to frost damage is dependent not only upon their intrinsic properties but also very much upon roof design. BS 473 [8] includes a permeability test which provides a measure of frost resistance for concrete tiles. A water absorption test is an implicit measure in BS 402 [9] which deals with clay roofing tiles. This Standard also calls for the rejection of tiles in which particles of lime are visible for, when wetted, these can cause local disruption of the tile.

3.4 CEMENT AND CONCRETE

Ordinary Portland cement is a material of major importance in construction. It is formed by burning ground limestone with clay and forms the basis of most concretes. As is well known, the cement sets and develops strength when mixed with water. The reactions which lead to this setting and strengthening are complex and are affected particularly by differences in manufacturing procedures and when admixtures are subsequently used. The chemistry of cement and concrete is treated comprehensively in specialist books and is not dealt with here. Other special cements are also made, with specific properties which are indicated broadly by their names, for example, sulphate-resisting, low-heat, rapid-hardening, white, coloured, hydrophobic. Cements are also made from blast-furnace slag produced during iron manufacture and by the addition of pozzolanic material (usually pulverized fuel-ash) to ordinary Portland cement. When limestone and bauxite are used during manufacture instead of limestone and clay, high-alumina cement is produced, which is characterized by a rapid rate of strength development and, if not 'converted' (see below), has a marked resistance to attack by sulphates. Cement is not used on its own in construction. It is the durability and properties of the concrete, mortar or rendering in which the cement is used which are of main concern and the effects of external agencies on the durability of concrete are now considered.

Concrete which becomes wet and then dries, alternately expands and contracts. The movement associated with wetting and drying is small, of the order of 0.04% for normal dense concrete; it is attributed to the cement matrix. However, some aggregates in the UK and, particularly, in Scotland, can themselves expand and contract on wetting and drying. Principally, these are igneous rocks of the basalt and dolerite types, and some sedimentary mudstones and greywackes. Moisture movement of concrete made with such aggregates may be at least doubled – and more than doubled if calcium chloride is used as an admixture. Movement around 0.08 to 0.1% may not seem large but it can lead to cracking of precast reinforced concrete units and to greater deflections of simply supported slabs. The use of such aggregates can also lead to a greater loss of pre-stress in pre-stressed concrete. It should be noted that the problem is mainly a Scottish one and most natural aggregates used in the UK do not show moisture movement of any significance. Design guidance on the shrinkage likely with these shrinkable aggregates, and the corresponding restrictions on usage, are given in Table 1 of BRE Digest 35 [10].

Some forms of silica found in aggregates can, in the presence of water, react with alkalis derived from cement and, in so doing, may cause expansion and subsequent damage to the concrete. The reaction is known as alkali/aggregate reaction. The problem, well known in the USA, seemed to be absent in the UK but a few cases have been reported in recent years from Jersey and the West Country affecting a dam in the former case and the concrete bases of electricity sub-stations in the latter. Since 1976, further trouble has occurred in the South-West and also in the Midlands, though it should be noted that the total amount of concrete affected is small. Several independent factors need to be present simultaneously before alkali/aggregate reaction can occur – the presence of reactive silica in the aggregate, sufficient sodium and potassium hydroxide released during the hydration of the cement and the continued presence of water. The reactive silica leads to the formation of a gel which absorbs water to give a volume expansion which can cause disruption of the concrete. The risk of damage is slight with most aggregates and cement used but reactive minerals have been found, particularly, though not exclusively, in sands and gravels dredged off the South Coast, the Bristol Channel and the Thames estuary.

The first signs of trouble take the form of fine random cracking and, when the alkali/silica reaction is severe, the presence of gel bordering the cracks which can be visible to the naked eye. The development of the reaction is slow. If the concrete will not be exposed to external moisture the likelihood of attack is negligible. In other cases much will depend upon the alkali content of the cement and upon the aggregate used and reference should be made to the specialist literature [11]. Unfortunately, tests of a specialist nature outside the scope of this book are needed to detect whether or not an aggregate is likely to be reactive.

A similar reaction can occur with some argillaceous dolomitic limestones known as alkali/carbonate reaction but no cases are known as yet in the UK.

Water freezing within the pores of concrete can cause disruption. Susceptibility to such attack is greatest with poor-quality concrete used in wholly exposed positions, such as kerbs and bridges. Good-quality, low-permeability concrete, used in most building situations, is not affected. Frost resistance can be increased greatly through the judicious use of air-entraining agents.

Water-soluble sulphates occur in some soils, notably the London,

Oxford and Kimmeridge Clays, the Lower Lias and Keuper Marl. They may be present, too, in materials used as hardcore, for example, plastered brick rubble. These sulphates of calcium, magnesium and sodium can attack the cement matrix to give reaction products which have an increased volume, and thus cause expansion. This, in turn, can lead not only to spalling and surface scaling but also to more serious disintegration. The extent of damage will depend greatly upon the amount and types of sulphate present, the ground-water conditions and the quality of the concrete. Once again, poor-quality concrete will be affected more drastically than well-compacted concrete of low permeability. Considerable resistance to sulphate attack can be obtained by using sulphate-resisting Portland cement to BS 4027 [12].

Steel reinforcement in concrete is inhibited from rusting by the high alkalinity of the surrounding concrete. Carbon dioxide, always present in the atmosphere, and sulphur dioxide reduce this alkalinity by carbonating the alkalis and thus increase the vulnerability of the steel to corrosion. In good-quality dense concrete, penetration by carbon dioxide is extremely slow and, when concrete cover is adequate, rusting may be almost indefinitely postponed. When concrete is permeable, or when cover is inadequate, carbonation can reach the depth of the reinforcement far more rapidly, with the consequent loss of rust inhibition. Corrosion of reinforcement or of pre-stressing steel can also occur if calcium chloride is used as an admixture, or if chloride occurs naturally in the aggregates used, and this can happen whether or not carbonation has taken place in the surrounding concrete. There have been many failures due to the use of calcium chloride, even when the total amount used was acceptable in the British Codes of Practice current at the time. Unfortunately, distribution within the mix can be uneven and harmful concentrations may occur. The problems of corrosion caused by the introduction of chlorides during mixing of the concrete are now more clearly recognized and restrictions are included in BS 8110 [13].

Rusting of reinforcement and subsequent spalling, cracking and general distintegration of concrete are a hazard near coasts, where attack by sea spray can occur, and steel needs to have a good dense cover of concrete, at least 50 mm thick. Even unreinforced concrete can distintegrate if attacked heavily by sea salts. The corrosion of reinforcement will lead generally to cracking of the concrete in the direction of the reinforcement, the position of which can be determined by means of a cover meter. Rust stains near such cracks will also often be seen. Rust staining, however, may sometimes appear not through cor-

Fig. 3.2 Corrosion of reinforcement causing cracking and spalling of concrete.

rosion of the reinforcement but because of the presence of pyrite in the aggregate. The resistance of reinforced concrete to attack can be increased by using austenitic steels as the reinforcement (see Section 3.5), particularly where cover to the steel is lower than desirable as in, for example, cladding panels of thin section. Epoxy resin coatings to the reinforcement are also likely to improve resistance to corrosion, but only if control of the coating composition and its application are high. It is most important to avoid breaks in the coating system.

Concrete may also be made with lightweight aggregates, such as clinker, expanded clay or slate, foamed blast-furnace slag and sintered pulverized fuel ash. The drying shrinkage of structural concrete made with these aggregates is about double that of concrete made with gravel aggregate and they are thus more liable to shrinkage cracking. Being more porous, lightweight concrete also offers less resistance to rusting of any steel reinforcement used. Some very lightweight concretes can be made using exfoliated vermiculite and expanded perlite as aggregates: drying shrinkage can be as high as 0.35%. Such concretes are not used structurally, however, but for fire-resistance and thermal-insulation purposes.

High-alumina cement in concrete converts, as a result of a change in the structure of the hydrated cement from an initial, metastable, form to a more stable form which possesses, unfortunately, higher porosity and lower strength. The rate at which this change and weakening occurs depends upon both ambient temperature and humidity. At one time it was believed that conversion was very slow at the temperatures and moisture conditions which could be expected in normal buildings. However, failures at a girls' school in Camden in 1973 [14] and at a boys' school in Stepney the following year [15] led to the conclusion that it would be advisable to assume that all high-alumina concrete would reach a high level of conversion at some time in its life. Highly converted high-alumina concrete in the presence of water is vulnerable to chemical attack such as, for example, might be caused through contact with damp gypsum plaster. Under such circumstances, very low strengths are likely to be reached.

3.5 METALS

Aluminium, copper, lead, steel and zinc are the metals most commonly used in building. The choice of a metal, or an alloy, for a particular design situation, is dealt with extensively in the technical literature and in the large number of British Standards and Codes of Practice which govern their use. The most important agencies affecting performance are those which have a primary influence on corrosion – these are moisture, and gaseous, solid and liquid pollutants.

Under completely dry conditions, corrosion does not take place but, in most situations in buildings, moisture is present and corrosion is a potential risk. It is wise to assume that moisture will be present at some time or other. Moisture acts as an electrolyte and, often, a galvanic cell is formed which leads to the loss of metal forming the anode of that cell. The most common example of this situation arises when two different metals are in aqueous contact with one another. However, galvanic cells can be created when some metals are in moist contact with other building materials, such as bricks or plaster; when an alloy has two or more phases with different electrochemical characteristics (as may occur with some brasses); and even in single metals when a difference in oxygen concentration may occur at the surface, for example in pitted steel where the base of the pit has less access to oxygen than the metal surrounding the pit. Galvanic cells may also be created, and subsequent corrosion occur, when particles of a metal or other substance are transported and deposited on other metals, as may happen in some plumbing systems. Corrosion generally is a complex electrochemical reaction which can be affected by the presence of dissolved atmospheric gaseous pollutants, by dirt, by manufactured admixtures (in particular, calcium chloride) and by temperature. Indeed, at some temperatures, bimetallic corrosion reactions may be reversed. The sensitivities to corrosion of the metals commonly used in building are described below.

3.5.1 Aluminium

Aluminium is used mainly for cladding, flashings and window frames. Aluminium in contact with copper and its alloys is readily attacked, and severe damage can be caused. This can happen even at a distance, and examples are known of aluminium flashings and window frames suffering severe pitting corrosion through rainwater draining from copper-covered roofs and depositing small particles of copper on the aluminium. Similar attack can occur from lead. It is essential to use the correct aluminium alloy for a particular building purpose. Many prefabricated aluminium dwellings built in the 1950s used a high strength aluminium–copper–magnesium alloy. The presence of the copper, together with damp conditions caused by heavy and repeated condensation, led to serious corrosion. Wood preserved with copper-containing preservatives can also attack aluminium in contact with it

Fig. 3.3 Corrosion of aluminium downpipe by copper salts.

and oak, too, can cause corrosion. Unprotected aluminium should not be embedded in cement mortars or in concrete, and direct exposure to sea spray will also lead to corrosion. Aluminium is not attacked by zinc, galvanized steel or stainless steel. Aluminium can be anodized, by treating it electrochemically to thicken the natural oxide film which forms upon it during normal atmospheric exposure. This thickened film confers upon the metal an improved appearance and resistance to corrosion. The anodized layer is readily disfigured by splashes of cement or lime and needs to be well protected on site: many cases are known of badly marked aluminium cladding and window fitments from that cause.

3.5.2 Copper

The main uses of copper are for roofing, cladding and plumbing. Copper is very resistant to corrosion and, when used in mixed metal systems likely to be encountered in building, it forms the non-corroding cathode. It can be attacked by flue gases containing sulphur dioxide if in close proximity to the chimney but this can be readily prevented by good chimney design or by the use of a copper–silicon alloy. Copper is corroded by ammonia and by some mineral acids but the likelihood of such a combination of circumstances is small in building. However, it is well to keep copper from direct contact with latex cements used for fixing some types of flooring. Waters with a high carbon-dioxide content dissolve copper but the rate of loss is small and the effects on the copper are slight. Some years ago, copper tubing used for water supply suffered serious pitting corrosion. Carbon derived from the lubricant used in tube manufacture formed a film which was nearly continuous and was cathodic to the copper surface. Pitting corrosion occurred at slight breaks in the film and caused rapid perforation of the tube. Nowadays, mechanical cleaning of tubes has largely eliminated this as a problem and tube conforming to BS 2871 [16] should be satisfactory. The risk may still exist when imported tube has to be used in times of shortage of British tubing.

3.5.3 Lead

Lead is used mainly for roofing (sometimes as a covering to steel sheets), for flashings and DPCs. It is highly resistant to corrosion through the formation of a dense, protective film of basic lead carbonate or sulphate. It is not, however, wholly immune from attack, and organic acids released from damp oak, Douglas fir and Western red cedar can cause corrosion: direct contact with these materials should be avoided. Lead is also attacked by free alkali present in rich cement/sand mortars, and severe corrosion has been known to occur within ten years with a DPC built-in with such a mortar. It is inadvisable to use unprotected lead sheet in such circumstances. Corrosion can be prevented by coating both sides with a bituminous paint.

3.5.4 Steel

Steel is used in building for structural purposes, either on its own or as reinforcement in concrete; for window frames; for cladding and decking; for rainwater goods; and in plumbing systems. Ordinary mild steel, that is, steel with no substantial amounts of alloying metals, rusts when exposed to the atmosphere. Mild steel is seldom directly exposed but has a protective coating of paint, bitumen or zinc, or is enclosed by other materials, usually by concrete, which reduces the rate of rusting. The effects of the agencies of deterioration upon steel used as reinforcement in concrete have been considered already. Steel protected by zinc, that is, galvanized steel, has a resistance to corrosion largely dependent upon that of the zinc (see Section 3.5.5). If steel is alloyed with small amounts of copper, about 0.25%, the corrosion rate in air is roughly halved and the weathering appearance is improved, for the corrosion product takes the form of a more tenacious coating than ordinary flaky rust. A greater resistance to corrosion can be imparted when steel is alloyed with at least 10% chromium, together with one or more other alloying elements. Steels so alloyed are known as stainless steels and the two main types used in building are ferritic and austenitic.

Ferritic stainless steels contain no appreciable amounts of nickel and some in the range have only modest resistance to corrosion. A ferritic stainless steel with 17% chromium is used in the building industry but experience of its behaviour is, at present, limited. Austenitic stainless steels contain nickel as well as chromium and the two most commonly used in building have a nominal composition of 18–8 (18% chromium, 8% nickel) and 18–10–3 (18% chromium, 10% nickel and 3% molybdenum). The former has only modest resistance to corrosion and is suitable, principally, for internal use: exposed externally to moist, polluted air, pitting corrosion is likely. For cladding purposes where good appearance is required, the 18–10–3 type is needed. Austenitic stainless steels are not affected by contact with most building materials and do not suffer in mixed metal systems, though they will act as the cathode in a mild steel/austenitic steel system and accelerate corrosion of the mild steel. An austenitic stainless steel is now used in domestic plumbing and heating.

One problem that has arisen is corrosion under stress. Failures have occurred with stainless-steel back boilers in hard-water areas. Though the full sequence of reaction is unclear, it is known that chlorides contained in many water supplies can concentrate behind the hard-water scale deposited in back boilers or in the air space when a boiler is not completely filled. At temperatures of around 80°C, cracking, known as stress corrosion cracking, may then occur – with, of course, failure of the boiler.

3.5.5 Zinc

Zinc is used in building for roofing, cladding, flashings and, sometimes, for rainwater goods. Additionally, much of the total zinc usage is in the form of protective coatings to steel. These may be applied by several techniques, of which hot-dipped galvanizing is the most common. Both hot dipping and metal spraying can achieve thick coatings at reasonable cost. The corrosion rate of zinc in unpolluted dry conditions is very slow and of no practical significance in building. In unpolluted damp conditions the rate may be four times as great and nearly ten times as great in damp polluted environments such as when zinc is exposed to sea spray or to sulphur gases or compounds. Zinc can be attacked by some wood preservatives and by inorganic flame-retardants. Rapid corrosion of galvanized wall ties has occurred when these have been embedded in black ash mortar, a source of sulphur compounds. Zinc is slightly attacked by wet concrete but progressive attack is inhibited by the formation of a calcium zincate which is insoluble in the prevailing alkaline conditions. This resistance to attack has enabled galvanized steel to be used as reinforcement in concrete when enhanced protection is required, for example, when difficulties are expected in providing the full depth of concrete cover normally expected. It should be noted, however, that bond strength can be reduced by hydrogen bubbles liberated in the initial reaction between zinc and alkalis in the concrete. To prevent this reduction, it is desirable to use galvanized reinforcement which has been 'passivated' by a chromate treatment.

Galvanized steel is also used for cold-water cisterns. Soft water, which contains dissolved oxygen and carbon dioxide, attacks zinc which, under such circumstances, will need protection by, for example, two coats of bituminous paint. Severe pitting corrosion has occurred when copper and brass debris from plumbing installation activities have been left inside unprotected galvanized cisterns and this should always be prevented. Zinc alloyed with copper can provide a range of brasses and one such, known as alpha-beta brass, is commonly used for fittings in water services. If the water supply is acidic, or is an alkaline water with a high chloride content, zinc can be removed from the brass (dezincification) leaving behind a spongy copper residue. This can lead either to penetration of the wall of the tap or fitting, allowing water seepage, or to their blockage by the zinc corrosion products. If such water supplies cannot be treated to increase their temporary hardness, which will prevent the attack, the use of hot-pressed alpha-beta brass fittings should be avoided.

Zinc cladding, roofing and flashings can be corroded by oak, Douglas fir and Western red cedar, and direct contact should be avoided.

3.5.6 Fatigue and creep

While corrosion remains the principal cause of deterioration of metals, some problems of fatigue and creep are worth noting. Metals subjected to a steady load deform slowly: movement depends upon load, temperature and time. This is known as creep and is not, in general, serious. Failures of lead flashings have occurred, however, partly as a result of the low resistance of lead to creep but more through its high thermal movement and low resistance to fatigue. This tendency can be inhibited by good design, with appropriate limitations on the size of any one piece of sheet lead used or by the use of lead containing a small proportion of copper. Technical recommendations are incorporated by the Lead Development Association in their technical guides. A zinc/titanium alloy is also available which has a greater resistance to creep than unalloyed zinc.

3.6 GLASS

Glass is a very durable material and is seldom affected by any of the agencies of deterioration mentioned. Surface etching can occur if sheets are closely stacked under damp conditions, and if alkali from paint removers splashes on to glass and is not removed. Thermal stresses can cause cracking, chiefly because glass has a rather different coefficient of thermal expansion than common framing materials. Proper allowance must be made in design and construction to accommodate differential movement and this aspect is dealt with in more detail in the section on cladding.

3.7 MORTARS AND RENDERINGS

Most present-day mortars consist of cement and sand – with or without an air-entraining agent – or cement, lime and sand. Mix proportions depend upon the types of brick and block used and their exposure. The effects of shrinkage caused by drying and by carbonation are generally insignificant. Mortars can be affected by frost but their resistance can be increased through the use of an air-entraining agent.

Serious distintegration of mortar can occur when soluble sulphates, sometimes present in wet brickwork, react with tricalcium aluminate, present in ordinary Portland cement mortars. A considerable increase in the volume of the mortar can then occur which causes the mortar to split and to become friable (see Chapter 7). Similar sulphoaluminate attack can occur when mortar is exposed to condensed water vapour containing sulphates derived from flue gases. This was a particular hazard when slow-combustion fuel appliances were used with unlined chimneys.

Typical external renderings are broadly similar in composition to mortars and have similar properties. Sulphate attack can occur through salts derived from wet brickwork: frost attack is rare. Drying shrinkage can be more of a problem and strong cement/sand mixes can craze and crack, particularly if applied in warm, dry weather.

3.8 PLASTICS

A wide range of plastics is used in building. PVC has the widest application, in either a plasticized or unplasticized form. Plasticized PVC is extensively used as a floor covering, as a sarking under pitched roofs, as a membrane for covering flat roofs and in the manufacture of water stops in concrete structures. Rigid unplasticized PVC is used principally for domestic soil and vent systems, for rainwater disposal and drainage, for wall cladding, as translucent or opaque corrugated roof sheeting and for ducting, skirtings and architraves. Rigid PVC is also used for window frames, sometimes in combination with metal or timber. Expanded PVC is available in board form for thermal insulation.

Polyester resins, reinforced with glass fibre, are used principally as cladding sheets but also for cold-water cisterns, in water and sewage disposal systems and for industrial gutters. Polyethylene, too, is used for cold-water cisterns and floats, for domestic cold-water pipes and for bath, basin and sink wastes. In sheet form, it is used as a damp-proof membrane and for covering concrete and hardcore surfaces. Acrylic resins are used for sinks, drainers and baths, for corrugated sheeting and for roof lights. Other plastics employed significantly in the building industry include acrylonitrile butadiene styrene copolymers for large drainage chambers; polypropylene for plumbing and drainage fittings and as wall ties; polycarbonates for glazing; phenol formaldehyde resins used to impregnate paper and fabric to provide wall and roof sheets, and foamed to give a cellular material used for thermal insulation; epoxide resins for *in situ* flooring and for concrete repair; and polystyrene, polyurethane and urea-formaldehyde in expanded form for thermal insulation. This is not a complete list, for many other plastics can find some use in the building industry, for example, nylon, cellulose acetate and polyacetals in taps and miscellaneous fittings.

Short-wave solar radiation degrades plastics by causing embrittlement and a change in surface appearance. The risk is greatest in coastal and rural areas. The effects can be reduced or increased by additives incorporated into the plastic. The addition of fire retardants reduces resistance to degradation. On the evidence so far available, the durability of plastics seems to be good. Moisture in general has little effect but can reduce bond strength between glass fibre and polyester resin: the extent of any weakening depends greatly upon the control exercised in manufacture. Plastics are not harmed, in general, by contact with other building materials, though cracking of polyethylene cold-water cisterns has been caused by the use of oil-based jointing compounds.

As will be seen from Table 2.2, PVC and polycarbonates have high thermal expansion. Unless properly allowed for, the movement of PVC gutters and down pipes can cause joint failure and leakage. Varieties of polyethylene have even greater thermal movement. Plastics creep under continued loads and special precautions are needed when stresses are high, as in filled cold-water cisterns – for example, by complete support of the base of the cistern.

The long-term behaviour of many of the plastics used is to some extent uncertain, for related experience is not yet available. More work is needed to enable better prediction of fatigue and creep under long-term loading. However, external performance so far has been generally good and for plastics used internally, or otherwise shielded from direct sunlight, the effects of short-wave solar radiation are nullified.

3.9 NATURAL STONE

Natural stones are classified as belonging to one of three main groups –igneous, sedimentary or metamorphic. Igneous rocks are formed by the solidification of a molten magma: common examples are granite, dolerite, basalt and pumice. The only one used to any extent in massive form in building is granite, though many may be used in crushed form as aggregates. Granite is highly resistant to all the agencies of deterioration described in Chapter 2: thermal movement is unexceptional and there is no moisture movement.

Sedimentary rocks may derive from particles produced from older rocks by the normal processes of weathering, from the accumulation of skeletons (usually marine organisms) and by chemical deposition. The particles forming the rock are deposited as sediments, through the action of water and wind, and are cemented together to varying degrees by minerals. Consolidation is assisted by pressure arising from the mass of the sediments as they build up in thickness. Sedimentary rocks have a natural bed though this may not always be apparent. Limestone and sandstone are the principal examples of sedimentary rocks and provide most of the building stone used in the UK. Most limestones consist essentially of calcium carbonate but a proportion of magnesium carbonate may also be present and, when this is significant, the limestones may be called magnesian limestones. Sandstones consist essentially of grains of quartz cemented together by silica (siliceous sandstones), calcium carbonate (calcareous sandstones) or, sometimes, by both calcium carbonate and magnesium carbonate (dolomitic sandstones). Sandstones containing appreciable quantities of oxides of iron are termed ferruginous.

The main cause of the deterioration of limestones and sandstones is atmospheric pollution. Sulphur gases dissolved in rainwater react with calcium carbonate to form calcium sulphate. When this crystallizes under dry conditions, it generates a stress which can break off particles of stone of varying size. If wetting is frequent from rainfall, the surface of the stone is slowly eroded and the calcium sulphate is continually removed. Under sheltered conditions, the calcium sulphate, which is only sparingly soluble, may build up to form a hard skin which causes more unsightly damage. When the skin eventually blisters and breaks off, it pulls away limestone with it. Dissolved sulphur gases can also remove the bonding calcium carbonate in calcareous sandstones and, in so doing, severely weaken the stone. Magnesium carbonate is also attacked by sulphur gases leading to similar deterioration of magnesian limestones and sandstones. Ground salts, and those in sea spray, can also cause damage. The former can cause expansive damage when penetrating into limestones and calcareous sandstones. Attack from sea spray is usually manifested by a general powdering of the external surfaces. Frost may attack some limestones, though rarely between eaves and DPC level: British sandstones are virtually immune to attack. The resistance of both limestones and sandstones to attack by frost, and to the crystallization of salts, depends essentially upon the pore structure of the stone. Resistance depends, particularly, upon the proportion of fine pores present and, to a lesser extent, upon total porosity. As the proportion of fine pores present increases, resistance to attack decreases, but this is a guide rather than a rule, and experience of use or specialist advice are the best aids to specification. A recent publication suggests zones within a building, for example, steps, copings, plinths and plain walling, within which the more common British limestones can safely be used given a particular environment [17].

Fig. 3.4 Contour scaling of sandstone.

A phenomenon known as contour scaling can cause damage to sandstones. Calcium sulphate can be deposited within the pores of sandstone, even though the latter may not contain calcium carbonate. The calcium sulphate may derive from limestones attacked by sulphur gases which has then washed on to the sandstone surface, or has been formed from some other source of calcium. Whatever its derivation, internal stresses are created in the surface crust and these are thought to arise from differences in thermal expansion and/or moisture movement between the sandstone and the crust blocked with calcium sulphate. The crust breaks away, usually at a depth of between 5 and 20 mm, and follows the contours of the surface [18].

Any deterioration through pollution, salts and frost is likely to be enhanced if sedimentary stones are laid so that the natural bed lies parallel to the vertical face of the wall. This is known as face bedding and should be avoided. The greatest durability is achieved when the natural bedding plane lies parallel to the horizontal courses in a wall.

Metamorphic rocks are derived from sedimentary rocks through the action upon them of great heat and pressure arising from movements of the Earth's crust. The structure of the rock is then changed radically. The only metamorphic rocks of significance for building purposes in the UK are marble (derived from limestone) and slate (derived from clays). Marble is used for cladding and is very durable: it is not immune from attack by sulphur gases but this is rare in reality. Slates are used for roofing, for cladding and as DPCs. Roofing slates are exposed to the most severe conditions and can be affected by sulphur gases. Those slates conforming to BS 680 [19], however, are highly resistant to attack. The Standard includes a sulphuric acid immersion test which is intended to simulate, but accelerate, the attack by sulphur gases dissolved in rainwater on any calcium carbonate present in slates.

One major cause of damage to all stones can arise if the fixings embedded in them corrode. In the past, the rusting of iron and steel cramps and dowels has caused extensive damage, particularly to limestones and sandstones. Appropriate metals to use are stated in BS 5390 [20].

3.10 TIMBER

There are many species of wood available which can be used for building purposes. Extensive technical data on the types, uses and properties of different woods are published in standard reference books and, notably, in the publications of the Princes Risborough Laboratory and the

Timber Research and Development Association. Softwoods are used for most structural and joinery work, and hardwoods for more specialized purposes, such as decorative flooring. The terms 'softwood' and 'hardwood' are not of particular significance in building and should not be taken as relating directly to strength and hardness. Most of the softwoods used belong to the species *Pinus sylvestris* (Scots pine, Baltic redwood), *Picea abies* (Norway spruce, European spruce, Baltic whitewood) and *Pseudotsuga menziesii* (Douglas fir, Columbian pine, Oregon pine). Larch, Western hemlock and Western red cedar are notable amongst other softwoods used.

The agencies which bear most upon the performance of timber are moulds, fungi, insects and moisture. The principal effects of these on timber are dealt with more conveniently in Chapter 2 and require little further comment. While sapwood is more susceptible to attack than heartwood it should be noted that even the heartwood of redwood, whitewood and Western hemlock is not naturally durable. Moreover market economics result in the presence of sapwood in most softwood available. Suitable preservative treatments are needed where there is any risk of attack by the agencies mentioned.

When seasoned timber absorbs and loses moisture, it expands and contracts and these moisture movements are much higher than for any other building material. Moisture movement in timber is a complex phenomenon and is affected by many factors, which include the speed of drying and wetting, the temperature at which the timber is seasoned, the direction in which the movement occurs and, of course, the species of timber. Tangential movement, that is, movement parallel to the annual rings, is greatest, followed by radial movement, movement at right-angles to the annual rings. Longitudinal movement in the length of the timber is smallest and, generally, disregarded. A standard method of measurement of moisture movement is to record the dimensional change which occurs when timber conditioned to equilibrium at 90% rh (relative humidity) and 25°C is dried to equilibrium at 60% rh and 25°C [21]. The movement is expressed as a percentage of the width of the timber specimen in this latter condition. Some typical tangential and radial movements for timbers commonly used in building are given in Table 3.1.

Timber subjected to changes in moisture can deform by bowing, twisting or cupping, and may crack if movements are frequent. Total moisture movement is not a sound guide, however, to the likelihood of distortion, for other factors can influence this, for example, grain direction and the presence of knots. To reduce the risk of distortion,

Table 3.1 Wetting/drying movements (approximately reversible)

Building material	Approximate movement (%)	
Asphalt, bitumen	Nil or negligible	
Glass	Nil or negligible	
Granite	Nil or negligible	
Metals	Nil or negligible	
Plastics	Nil or negligible	
Clay bricks	0.02	
Calcium silicate bricks	0.02–0.04	
Concrete bricks	0.04–0.06	
Dense concrete	0.03–0.04	
Structural lightweight aggregate concrete	0.03–0.08	
Concrete with shrinkable aggregates	0.05–0.09	
Limestone	0.06–0.08	
Sandstone	0.06–0.08	
European spruce, Baltic whitewood	1.5*	0.7†
European larch	1.7*	0.8†
Douglas fir, Oregon pine	1.5*	1.2†
Western hemlock	1.9*	0.9†
Scots pine, Baltic redwood	2.2*	1.0†
English oak	2.5*	1.5†

*   Tangential
†   Radial

good practice requires that timber is seasoned to around the average moisture conditions likely to be met in its place of use. BS 5268 Part 2 [22] relates these moisture contents to positions within a structure.

Timber is an organic material and, as such, is affected by prolonged exposure to sunlight. Short-wave solar radiation causes degradation of the surface and timber stored on building sites unprotected from rain and sun acquires a typical grey appearance. Some fungi can discolour wood without affecting its general durability and strength. One such fungus is called blue-stain, which can stain undesirably the sapwood of some softwood. This is a problem encountered before use, however, for the fungus does not grow on timber used at moisture levels considered safe in buildings. If excessive moisture remains in any timber used, attack by far more serious fungi is likely to develop, as already described.

3.11 MOISTURE MOVEMENT

Many of the building materials mentioned in preceding paragraphs expand and contract when they take up or lose moisture. This movement is not always wholly reversible but usually approximately so. It depends, *inter alia*, upon the degree of wetting and drying, the precise composition of the material and the plane of measurement. Typical movements shown in Table 3.1 are those which could be caused by changes in moisture content likely to be encountered in environments common to normal buildings. This should be taken as a guide only and the specialist literature should be consulted if more precise values of movement under defined changes in the moisture regime are sought.

4

Moisture

The special position of moisture as the principal agent causing deterioration is mentioned in Chapter 2. Moisture cannot be totally excluded from buldings: a knowledge of the main sources of moisture, however, will enable steps to be planned to minimize the total amounts trapped, entering or generated in a building and so help reduce or prevent some of its detrimental influences.

4.1 WATER ENTERING DURING CONSTRUCTION

Much water is used during construction for the mixing of mortar, concrete and plaster and for wetting bricks before laying. For brickwork alone, as much as one tonne of water may be used in building the average house. Some of this constructional water is immobilized in the hydration of cement and plaster, and some evaporates before occupation of the building. Even so, much water necessarily used for mixing purposes will still be retained and can be slow to dry. Not a lot can be done about this moisture, though good ventilation and the judicious use of heating in the first year will assist drying. Unfortunately, a good deal of water is introduced quite unnecessarily by poor site control of the storage of materials and by inadequate protection of partially built structures. It is all too common to see timber roofing and joinery, blocks intended for internal use and stacks of bricks left wholly unprotected from the rain. These components are then installed wet and often left uncovered, so that further rain keeps them wet. Flat roof decks, especially those screeded with lightweight concrete, can absorb large quantities of rainwater if not adequately covered during the constructional phase. An average house, with masonry walls, is likely to contain several tonnes of water just after completion when site control is of a standard commonly seen today. At least a year is likely to elapse after occupation before the moisture level drops to that in equilibrium with normal internal humidity conditions. Efficient site storage and protection could go a good way towards reducing these amounts. In particular, the tops of walls, floors and roofs should be protected, as far as possible, and all timber completely protected from rain and ground water.

4.2 GROUND WATER

In the UK, the level of the water table is seldom far below the surface of the ground. Materials in contact with the ground will draw up this water by capillary action into their pores and into the structure of which they are a part. It should not be assumed that this will not happen, even on an apparently dry site. Building operations undertaken below ground can, in themselves, change, sometimes detrimentally, the pattern of natural water drainage and also the level of the water table. The height to which ground water can rise if not obstructed can be considerable: it is affected by a number of factors, chief of which are the pore structure of the materials in contact with the ground and the depth below ground of the water table. These ground waters contain salts in solution which tend to

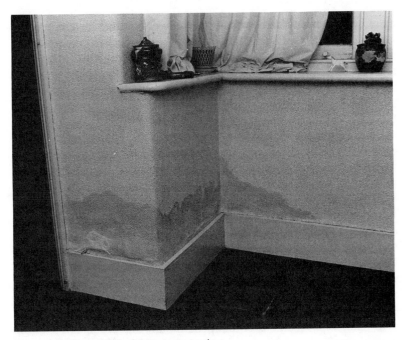

Fig. 4.1 Stains typical of damp penetration.

concentrate on wall surfaces. The salts are often hygroscopic and so add to general dampness by absorbing moisture from humid air. It might be thought that entry of ground water and its associated salts is a problem of the past – and so it should be, since the incorporation of a DPC was made compulsory in 1875. Unfortunately, cases of attack are not infrequent in new buildings and arise, generally, not because a DPC has been omitted but because it has been bridged and so made ineffective. Common causes of bridging are considered in the chapters dealing with floors and walls.

The passage of ground water into a structure should be suspected when internal wall surfaces show persistent dampness in an irregular pattern, to a height which can, on occasions, exceed 600 mm. The greatest heights reached occur on damp sites and when the outer faces of the external walls do not allow easy drying, for example, if covered with a non-porous coating. When salts are present, decorations may be pushed off. The salts are usually left behind and seen as a white powdery growth.

4.3 RAIN AND SNOW

From the earliest times, a primary purpose of any building has been to shelter the occupants from rain and snow, and it is disturbing that so many buildings fail to do so adequately. Rain and snow may penetrate directly through gaps in the structure, particularly at the junction of windows and doors with walls, at joints between cladding panels and at gaps in sarking felts. It may enter indirectly by passage through porous building materials under the action of capillary forces. Passage may be assisted by gravitational forces, as when water penetrates through a flat roof. Indirect penetration commonly occurs through blocked cavity walls, through solid walls ineffectively rendered or pointed, and particularly through flat concrete roofs.

The risk of rain penetration through gaps in a structure is greater when the rain is wind-assisted than in still conditions. The severity of exposure to rain was seen as likely to vary both with rainfall and with wind speed: measurements made by Lacy, using rain gauges set in the walls of buildings, showed that the amount of rain driven on to a wall was directly proportional to the product of the rainfall on the ground and the wind speed during the rain. This relationship was used as the basis for the derivation of an index of driving rain which could be used to give relative probabilities of rain penetration. The product of average annual total rainfall in metres and annual wind speed in metres per second provides the annual mean index of exposure to driving rain. Contour maps have been produced showing the annual mean driving-rain index for areas in the British Isles. The wind speed used in the calculation of the index refers to an open, level site, so local correction factors have been derived which allow for local topography, for the roughness of the terrain and for the height from the ground to the top of the building [23]. These factors are designed to allow local driving-rain index values to be obtained which relate to the amount of rain penetration through walls which actually occurs.

Computer analysis of meteorological data has, recently, enabled the Meteorological Office to produce more realistic values based on the fact that heavy rain is usually associated with strong winds. Two indices have consequently been derived. The local annual index measures total rainfall in a year and is most significant for the average moisture content of masonry. The local spell index measures maximum intensity in a given period and is most significant for rain penetration through masonry [24].

The amounts of rain blowing on to walls can be considerable: even in moderate conditions, up to one tonne per square metre a year can be so driven. The pattern of wetting of the walls of any individual building will vary with its fabric and the extent to which it controls absorption and run-off – and, of course, with individual design features (see Fig. 4.2). As a generality, it can be said that driving rain tends to concentrate at internal and external angles and, particularly, at the edges of buildings. The extent to which walls absorb rain depends much upon the pore structure of the materials used – there can be a difference of 100 to 1 between types of bricks. The rate of drying after wetting is greatly dependent upon pore structure and upon the ambient temperature, humidity and wind speed. The extent of drying depends on the frequency and duration of the dry spells between periods of rain. Joints between non-porous components are particularly vulnerable to penetration by rain or snow under wind pressure, for they are unable to hold moisture by absorption and so retard its passage to the interior. When rain penetrates a crack in a relatively impervious system, for example, in a dense rendering, it can be slow to evaporate and can cause many of the problems associated with rain penetration which might have been avoided had free evaporation been possible.

Falling snow has a density typically only one-tenth that of rain. It can, as a consequence, be blown upwards to a greater extent and also distributed unevenly on structures. It is not uncommon for it to be blown under eaves and sarking, and to build up so that, on melting,

Fig. 4.2 How not to dispose of rainwater.

considerable wetting of roof timbers takes place. Although it is not a major problem in the UK the effect of eccentricity of snow loads on pitched roofs needs to be consciously considered in design. Heavy snow accompanied by high winds and the subsequent drifting has caused partial or total collapse. Guidance on the high local loads that drifting snow can impose is given in a recent BRE Digest [25].

Direct penetration of rain or melted snow shows as damp patches, varying in size and position, on the inside face of external walls, usually within a few hours of the beginning of precipitation. These patches appear most often around window and external door frames, but not exclusively so. Competent building design and construction should prevent such penetration. Many of the traditional ways of keeping rain away from the main structure, for example, by the use of a good eaves overhang and by recessing windows, seem to have been abandoned in most modern dwellings, possibly through ignorance, indifference or false economy. They need to be reintroduced.

4.4 MOISTURE FROM HUMAN ACTIVITIES

A great deal of moisture is produced from normal human activities and this can be a major input to help cause condensation, which is considered in some detail in the next section. Just by normal breathing, one

Table 4.1 Typical daily moisture production within a five-person dwelling

Activity	Moisture emission (litres)
Clothes drying	5.0
Cooking	3.0
Paraffin heater	1.7
Two persons active for 16 h	1.7
Five persons asleep for 8 h	1.5
Bathing, dishwashing	1.0
Clothes washing	0.5
Total	14.4

Based on data contained in BS 5250:1975. (The above does not take into account any moisture removed by ventilation.)

person produces at least 0.3 litres of moisture in a day. Typical domestic activities can greatly exceed this amount. Clothes washing and drying constitute a major source of moisture input. Drying a normal wash for a family of five can generate as much as 5 litres of moisture, some ten times as much as is likely from the washing phase. To add to the difficulties, the drying of clothes indoors is likely to occur most often when the weather is damp. It is not uncommon, too, for large volumes of water to be left in baths and sinks, for considerable periods of time, to soak crockery and clothes. More than half the daily input of moisture in a typical home is produced in the kitchen.

Unvented heaters, such as paraffin stoves and free-standing gas appliances, including gas ovens, which are often used as supplementary heaters, generate large amounts of moisture. Paraffin produces more moisture than fuel consumed – approximately 1.2 litres per litre of paraffin. The use of such heaters has grown in recent years, as their fuel costs are usually lower than the cost of electricity. Table 4.1, based on data in BS 5250 [26], shows some moisture outputs from typical domestic situations and the pre-eminence of clothes-drying as a producer of moisture.

4.5 CONDENSATION

The presence of moisture may lead not only to direct attack on materials but also to condensation. It is probable that condensation and its effects have been the greatest single post-war problem in building, particularly in local-authority housing. Condensation and mould growth now affect some two and a half million homes in the UK. The cause of condensation is simple. Air at any time will contain some water vapour and warm air is able to hold more water vapour than cold air. When air containing moisture is cooled progressively, there will come a temperature at which the air cannot hold all the moisture any longer and it is then said to be saturated. This related temperature is known as the dew-point and is the temperature below which condensation will begin. Water vapour in air exerts a pressure and this pressure causes it to move through all except completely impermeable materials to areas of lower vapour pressure. At any given temperature, the ratio of the actual vapour pressure exerted to the vapour pressure which would be exerted if the air were saturated is known as the relative humidity (rh); the higher the rh, the closer the air is to saturation.

The relationships between vapour pressure, dew-point, rh, temperature and the actual amounts of water contained in air can be represented by a psychometric chart such as that shown in Fig. 4.3. This looks complicated but two examples will suffice to show that its use is simple. If the air within a room at a typical temperature of, say, 20°C has a relative humidity of 40% (point A) and this air is then cooled, reading horizontally to the left, point B indicates the dew-point (about 6.5°C) and, if any air reaches a part of the structure just below this temperature, then condensation will occur. The chart also shows that air at 20°C and 40% rh contains about 6 grammes of moisture in a kilogramme of dry air. If some internal activity adds a further 6 grammes say, by the drying of clothes – then the air will contain 12 grammes of moisture and be at a relative humidity of just over 80% (point C). A drop in temperature to 17°C will then suffice to allow condensation to take place (point D).

Any surfaces below the dew-point of the air immediately adjacent to them will suffer surface condensation. Visible condensation is seen frequently on windows, external walls and on cold pipes and cold-water cisterns, and gives a warning sign that moisture levels are high. As vapour pressure is usually higher inside a building than outside, water vapour will move through a permeable structure towards the outside. In its passage, it may be cooled to the dew-point, causing condensation at some point within the constructional materials, in the spaces between them or at cold bridges, which are localized cool surfaces interrupting areas of better thermal insulation, for example, lintels. Interstitial condensation, that is, condensation within the thickness of a material or structure, can be potentially harmful.

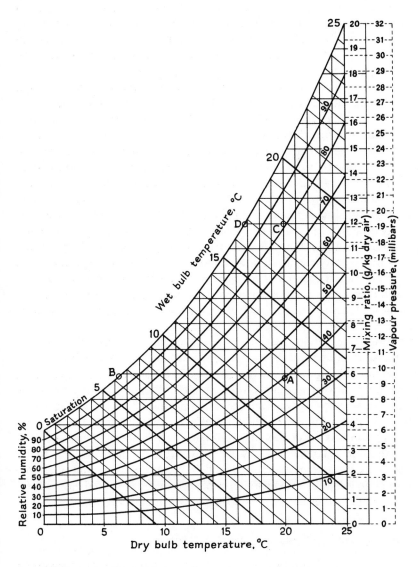

Fig. 4.3 Psychometric chart.

A main effect of excessive moisture reaching the inside of a building –whether arising from moisture trapped during construction, as ground water, as direct rain penetration or from human activities – is to increase the relative humidity and the risk of condensation, and assist the growth of fungi (see Fig. 4.4). Fungal growth can occur if the rh remains above 70% but, for active growth, prolonged spells of over 80% rh are generally necessary or continued access to direct moisture supply within the material upon which they will form. The spores of fungi are ever-present in the air in most buildings and so are nutrients; thus, the factor controlling growth is this supply of moisture. Fungi are usually dark in colour, often green or black, and both disfigure and cause deterioration of furnishings, fabrics, wallpapers and many decorative coatings. They can form on stored clothing and bed-linen, particularly when these are in unheated, unventilated built-in cupboards on external walls. Certain fungi can also rot timber and this is dealt with in Chapter 3. Suffice it to say here that heavy and repeated condensation running down windows and entering the timber frame through discontinuities in the paint film or in the putty can allow wet-rot fungi to develop. In the past, drained condensation channels were often provided, though condensation was not then a serious problem. Nowadays, when reduced ventilation rates and changes in living habits have made condensation a major problem, such drainage channels are rarities.

Moisture vapour reaching porous building materials will raise the moisture content of the air contained within the pores: if liquid water enters the material, it will directly raise its moisture content. The effect of both sorts of wetting is to increase the thermal conductivity of the material and so reduce its level of thermal insulation. If heating within the building remains constant, the effect will be to reduce the temperature and this reduction will itself help, still further, to increase condensation. (The effect may be, of course, to cause heating to be increased and to be manifested partly in greater fuel expenditure. If this heating is by unvented paraffin appliances or by unvented gas heaters, then still further amounts of moisture are generated.)

4.6 AVOIDANCE OF CONDENSATION

The amount of moisture available as a potential cause of condensation can be reduced by attention to the foregoing points. Thus, good design and construction can minimize the amounts of moisture entrapped during construction, entering from the ground and penetrating as rain or snow. Publicity and education may help in reducing the actual

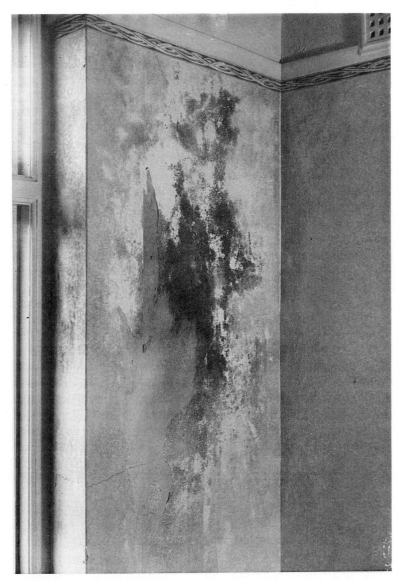

Fig. 4.4 Bad mould growth.

generation of moisture within a building, in promoting ventilation and increasing heating levels. It is probable, though, that little heed will be given to advice which results in higher fuel bills. Nevertheless, heating by oil and gas appliances which are not connected to flues should be actively discouraged. Mechanical extraction should be used to remove moisture from high emission areas, such as kitchens, bathrooms and drying cupboards. Electric dehumidifiers may have a role but they are not very effective in buildings where the main problem is low temperature rather than high vapour pressure. They are portable but rather noisy and will be more acceptable in kitchens than in bedrooms.

However, even when all such precautions are taken, it is inevitable that moisture will be present in buildings. Whether or not condensation will occur in a specific situation requires knowledge of many complex and interrelating factors. These include not only details of the design and the materials to be used but also knowledge of the occupations to be carried out in the building and their pattern. Slight changes in habits can tilt the balance towards or away from condensation. In particular, much depends upon the levels of heating and thermal insulation used, ventilation and the permeability of the building fabric to moisture. Heating a building helps to raise the internal surfaces above the dew-point of the air. The effect of increasing thermal insulation for any given heat input is to conserve the heat and thus make it more effective. The temperature of surfaces is also dependent upon the speed at which the building fabric heats or cools in response to changes in external and internal temperatures. Ventilation assists in the removal of any moisture produced within the building. Natural ventilation is nearly always wholly beneficial in reducing the risk of condensation: modern building design, with its tendency to dispense with chimneys and flues, and the current trend to block chimneys in existing buildings, has reduced these ventilation rates. The permeability of the construction influences the moisture conditions within the structure and can be changed greatly by the use of vapour barriers. The vapour resistance of some common membranes and materials is shown in Table 4.2.

BS 5250 recommends design assumptions for dwellings relating to internal and external temperatures, to humidities and to ventilation rates. It recommends standards of thermal insulation which, taken in conjunction with those design assumptions, should prevent surface condensation in dwellings in all but the most exceptional circumstances. The principles used can be applied to specific floor, wall and roof constructions to predict the likelihood of condensation and the point in the structure at which it could occur.

Table 4.2 Vapour resistance of materials

Material	Vapour resistance, approximate [(MN s)/g]
Aluminium foil	Several thousand
Polyethylene: 100 μm	250–350
Gloss paint	7–40
Roofing felts	4.5–100
Stone: 100 mm	15–45
Brickwork: 225 mm	5–30
Expanded polystyrene: 50 mm	5–30
Bitumen-impregnated paper	11
Concrete: 100 mm	3–10
Wood wool: 50 mm	0.7–2
Rendering: 20 mm	2
Foamed urea-formaldehyde: 50 mm	1–1.5
Plasterboard: 9.5 mm	0.4–0.6
Kraft paper: 5-ply	0.6
Emulsion paint: 2 coats	0.2–0.6
Plaster: 5 mm	0.3

Based on data contained in BS 5250 (1975).

Concern with failures has so far related mainly to condensation which occurs on the inside surfaces of a building. Clearly, when this is sufficiently severe to cause damage to fittings, furnishings and decorations, such an emphasis is understandable and inevitable. However, condensation on windows and on internal surfaces and its effects are, at least, readily seen. As the standards for thermal insulation rise, it is probable that the point at which condensation occurs will be pushed back into the structure, towards the outside, where it may be hidden but do greater damage in the long run. It may not be possible to prevent interstitial condensation but it is necessary to ensure that it does not occur where it can cause the types of damage described in Chapters 2 and 3, particularly to timber and to metal fixings. Detailed knowledge of the thermal properties and behaviour of materials and structures under fluctuating temperature, moisture and ventilation conditions is still a field for research: guidance is available in the specialist literature to enable the risk and position of condensation to be assessed for a specific structure. If past experience is any guide, though, it would be unwise to believe that the extent and position of condensation can be closely controlled or that vapour barriers will prevent, in reality, all water vapour from passing. It will be wiser to assume that condensation will occur at any point within a structure and to ensure, by suitable design and specification, that the moisture can move freely and easily to the outside, either in liquid or vapour form, and that sensitive materials are desensitized, for example, by preserving timber components, or replacing them by those products more resistant to moisture.

5
Foundations

It seems appropriate to start with foundations when dealing in detail with the avoidance of defects in specific building elements. That is where construction starts and, also, serious defects in foundations are, generally, the most difficult and costly to remedy.

The Building Regulations identify three functions for the foundations of a building. The first is that they should sustain and transmit safely to the ground the various loads imposed by, and upon, a building, in a way which will not impair the stability of, or cause damage to, the building or adjoining buildings; secondly, their construction must safeguard the building against damage by physical forces generated in the subsoil; and thirdly, they must resist adequately attack by chemical compounds present in the subsoil. Clearly, the ground or subsoil has a major influence on the ability of foundations to perform these functions and its type, structure and properties are of prime importance.

5.1 TYPE AND STRUCTURE OF THE SOIL

Soils include materials of various origins but, for purposes of identification, two essential characteristics have been recognized – the size and nature of the particles composing the soil and the properties resulting from their arrangement [27]. Five principal types can be distinguished. Gravels and sands are relatively coarse-grained, non-cohesive particles derived from the weathering of rocks. Gravels consist of particles which lie mostly between 76 and 2 mm: the particle size of sands lies between 2 and 0.06 mm. Within this range, sands may be classified as coarse, medium or fine. Silts are natural sediments of smaller particle size than sands, chiefly lying between 0.06 and 0.002 mm. Clays are formed from the weathering of rocks and have a particle size less than 0.002 mm. Peat is an accumulation of fibrous or spongy vegetable matter formed by the decay of plants. These five soil types and their characteristics are shown in Table 5.1.

Table 5.1 Classification and characteristics of common soils

Type of soil	Size and nature of particles	Compressibility
Gravels	Mostly between 60 and 2 mm; non-cohesive	Generally loose and uncompacted
Sands	Mostly between 2 and 0.06 mm; non-cohesive when dry; may contain varying amounts of silt and clay	Can be loose or somewhat compacted
Silts	Mostly between 0.06 and 0.002 mm; some cohesion; little plasticity	Capable of being moulded in the fingers
Clays	Less than 0.002 mm; marked cohesion and plasticity; appreciable shrinkage on drying and swelling on wetting	Very soft clays exude between the fingers when squeezed; firm clays can be moulded in the fingers by strong pressure; stiff clays cannot be so moulded
Peats	Fibrous organic materials; can be firm or spongy	Have a high degree of compressibility

5.2 INTERACTION BETWEEN SOILS AND BUILDINGS

Under natural conditions, water and air fill the spaces between the soil particles. The properties of soil are influenced greatly by the amounts of water so held, and the volume and strength changes which may occur when this water is reduced or increased. Changes in the behaviour of soils influence that of the foundations in contact with them and this affects the behaviour of the superimposed building. The building itself, through the loads transmitted via the foundations, compresses the soil and can change its behaviour. Interactions between the soil, the foundations and the building are complex and highly dependent upon the forces involved when soils shrink or expand due to loss or gain of moisture.

5.3 SOIL MOVEMENT

When the water present between soil particles is removed, the latter will tend to move closer together: conversely, when water is absorbed, they will tend to move apart. Large movement can occur with clays, for these are capable of absorbing and relinquishing large quantities of moisture: drying leads to shrinkage and a gain in strength, and absorption to swelling and a loss in strength. Movement in sands is for the most part negligible, for they have little capacity to hold water. Silts have movement which lies between that of clays and sands. Peat can exhibit very large movement and has little bearing capacity.

Changes in water content of soils may be caused in several ways. The most obvious is that caused when the soil is loaded by the weight of the foundations and the superimposed building. Water is then squeezed out of the soil and the soil particles move closer together. As the ground is compressed or consolidated in this way, the foundations settle, until equilibrium is achieved between the load imposed on the soil and the forces acting between its particles. The more clay there is contained in the soil, the longer does it take for this equilibrium to be achieved. With soils wholly of clay, such settlement may go on for years while, with sands, it is rapid and is substantially finished by the time building is completed. It may be of interest to note that a reduction in loading, such as will be caused by demolition or excavation, can lead to water migrating towards the unloaded soil, causing it to swell – again, appreciable with clays and negligible with sands.

5.3.1 Effects of vegetation

Knowledge that movement can be caused by loss of water through the growth of vegetation and to gain of water by its removal seems to have been overlooked – at least, up until 1976, when the severe drought and hot weather in the UK led to a rash of troubles. In fact, problems associated with vegetation and climate were of long standing, and both researched and reported upon by Ward [28] in the immediate post-war years. Tree roots can extract large quantities of water from soil: a fully grown poplar uses over 50 000 litres in a year. When the soil is of clay, this will lead to a drying shrinkage, the magnitude of which will depend upon the inherent properties of the clay and, of course, on the nature of the tree and its moisture requirements. If tree roots take up moisture from under, or near to, foundations, the latter will subside and such subsidence will almost inevitably be uneven. The possible adverse effects on foundations and, thus, upon the final structure, of the drying action of tree roots in areas of shrinkable clays is not appreciated by many involved in the construction process – and, unfortunately, by many owners of buildings. The small, immature tree, which seems to be a fair distance away initially, looks uncomfortably close in later years.

Table 5.2 Heights of some common trees

Common name	Approximate mature height (m)	Approximate height after 15 years (m)
Douglas fir	40	12
European larch	40	9
English elm	40	7
Lombardy poplar	30	9
European ash	30	7
London plane	30	5
White willow	25	15
Scots pine	25	10
Silver birch	25	9
European horsechestnut	25	9
Oak	25	8
Beech	25	7
Lime	25	7
Cedar	25	6
Weeping willow	20	9
English holly	15	4
Yew	10	5
Juniper	5	3

Based on data contained in *The International Book of Trees* (Mitchell Beazley, 1973).

The mature height of some common trees found in the UK is shown in Table 5.2. The distance to which the roots of a tree spread depends largely upon the type of tree and its height. The roots of many common trees extend to a distance at least equal to their height. The roots of willow, elm and poplar can extend to twice the height.

It is also most important to understand that, when trees are felled, clay soils will gradually swell as water returns to the ground. A clay site cleared of trees needs to be allowed to recover before building begins or, if this is not possible for economic or other reasons, then foundations need to be specially designed, as described later, to prevent damage caused by this swelling. Hedges and shrubs can also cause desiccation and shrinkage of clay and, if removed, can lead to swelling. While their effect is less than that produced by trees, it cannot be neglected, for they are often grown closer to buildings. The clays which display the largest movement upon drying are the firm, shrinkable clays found particularly in South-East England, for example, the London and Oxford clays, and the Weald and Gault clays.

5.3.2 Other causes of ground movement

Major ground movement can occur over underground mining areas, as the ground collapses over the workings. Nowadays, good records are kept of mine workings but this was not so during much of the 19th century.

A rather uncommon combination of circumstances can lead to the expansion of ground when frozen. The soils mainly involved are silts, fine sand and chalk. In areas where the water table is high and when there are prolonged periods of freezing, ice lenses can be formed in these soils which cause heaving of the ground and the foundations upon it – a phenomenon known as frost heave. In practice, little trouble has been caused in the UK and, even in severe winters, the ground at foundation base level near to occupied and heated buildings is not likely to freeze. The risk, small anyway, is confined to unheated buildings and those under construction.

5.3.3 Effects of foundation movement

The greatest problems have occurred when shrinkable soils have dried excessively through the removal of moisture by nearby growing vegetation. Such drying is likely to be greatest at the corners of foundations. As the ground falls away, the weight of the building pushes the then suspended parts of the foundations down and the walls in that vicinity crack. Cracking is predominantly diagonal and follows the vertical and horizontal mortar joints in brickwork, unless the mortar is abnormally strong for the bricks used, when cracking may occur through the latter. The cracks are widest at the top corners of the building and decrease as they approach ground level. The appearance of cracks of this pattern at the end of a specially dry summer is a fairly sure sign of desiccation of a shrinkable clay soil. Door and window frames also distort due to the deformation of the walls, leading to their sticking or jamming. In severe cases, service pipes may fracture, walls may bulge and floors may slope noticeably. The cracks tend to close partly, following periods of prolonged rain, for example, by the end of the following winter.

When trees, large shrubs and hedges are cut down before building, long-term swelling of clay soils can be substantial and can take place over several years. The upward forces on foundations can cause severe

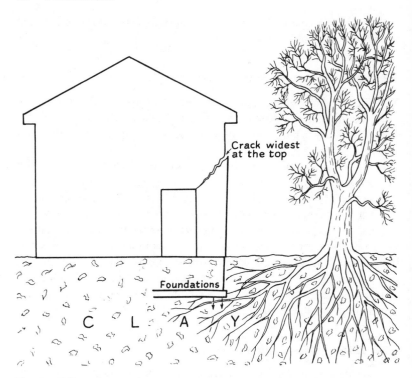

Fig. 5.1 Effect of tree roots on foundations. Tree roots cause clay to shrink and foundations to drop.

stresses at the corners of the building or may act more centrally. In the former case, cracking patterns are usually similar to those already mentioned but with the important difference that crack width is greatest near to foundation level and becomes narrower at the higher levels. When forces act more centrally, cracks tend to be straight rather than diagonal and are widest at the top. Often, there will be a single crack in each of the two opposite walls of the building and they may be connected with a crack in the floor if this is of concrete.

When subsidence occurs due to active mining operations, the building tilts towards the advancing working and random forms of cracking generally occur. Diagnosis is fairly obvious through knowledge of the presence of active mining in the area. Over old forgotten workings, diagnosis is clearly more difficult but the presence of random cracking is a guide, though this can occur from other causes, for example, from the use of poorly consolidated fill or from seismic tremors.

5.3.4 Avoidance of failure due to soil movement

Detailed guidance on the design and construction of foundations is given in BS CP 2004 [29] and BS CP 101 [30]. The latter deals with non-industrial buildings of not more than four storeys and it is in this category of buildings that most post-war problems have occurred, particularly in housing. The National House-Building Council (NHBC) has produced a related manual intended for house builders, engineers and architects [31] which gives a check-list of actions required to prevent trouble. A first need is for adequate site reconnaissance, and a study of any local recorded information, to determine the topography, the basic type of ground, the vegetation present, and the type and proximity of ground water, and to glean as much information as possible on the previous use of the site, for example, any possible relationship to mining. Geological maps will give guidance on the main ground strata to be expected but study of these needs to be augmented by the digging of trial pits (or hand-augered bore holes) and by descriptions in detail of the soil profiles revealed and the level of ground water. If such inspections reveal the likelihood of hazardous ground conditions, such as those associated with peat or shrinkable clay, or the presence of mining or of old buildings, then specialist advice needs to be sought. Failure to undertake these essentially simple steps has caused much expenditure in post-war years.

On sites where the soil is of firm shrinkable clay, it is necessary to take the foundation down to a depth which should eliminate significant ground movements. A foundation depth of 1 m is generally adequate in such circumstances, when the site is unaffected by trees. The NHBC has provided guidance on the depth of foundations required on sites with trees, relating such depth to the type of tree, its mature height, its distance from the foundations and its geographical location [32]. Several types of foundation can be used in these circumstances and choice is likely to be dictated by financial considerations. The traditional strip foundation of concrete is usually some 150 mm thick with a minimum width of 450 mm, from which two leaves of brickwork are built up to DPC level. At a foundation depth of 1 m, at least twelve courses of brickwork would be needed. The cavity wall up to ground level is filled with concrete and the trench is backfilled with earth and hardcore. These traditional strip foundations are labour-intensive and, as the depth required increases, trench fill or narrow strip foundations are being used increasingly. These are formed by cutting a trench narrower than that needed for the traditional strip and pouring concrete to a depth such that generally only four courses of brickwork are required to reach DPC level. It is important that setting-out is accurately accomplished to ensure that the brick courses do not oversail the edges of the narrower strip. As the depth for foundations increases, and in soils liable to swell, it can become unsatisfactory to use ever-deeper trench fill. Swelling pressures can act on the large areas of foundations then in contact with the soil to cause lateral movement and rotation of the foundations and also vertical movement on the sides of the trench fill. It will be better, and often cheaper, to use bored piles.

The likelihood of damage from the swelling of clay soils following the cutting or removal of vegetation is difficult to predict but it is known that many years can elapse before movement can be considered complete. BS CP 101 considers it unwise to assume that swelling will be completed within two winter seasons following the removal of trees and hedges, and calls for special foundations, such as bored piles and ground beams, to be used if buildings are sited over or close to former trees, shrubs or hedges. The time scale of swelling, and the distance over which the effect is appreciable, requires further research.

If additions are to be made to existing buildings in areas where soil movement is likely to be a problem, then it is probable that differential movement will occur between the new and old structure. It is necessary to ensure that such movement can take place without causing damage to either by the use, where possible, of flexible or sliding joints.

Where subsidence due to mining is a risk, expert advice needs to be sought on the design of foundations, which is outside the scope of this book. The National Coal Board, or the equivalent authority for other types of mining, should be consulted. Detailed guidance on the types of foundation for low-rise buildings, with particular reference to special ground problems, including soils of firm shrinkable clay and mining subsidence, is given by Tomlinson, Driscoll and Burland [33].

5.3.5 Remedial measures

Repairs to foundations are very expensive and, if things have gone wrong, much care should be taken in deciding whether repairs are necessary and, if so, the form they should take. There was a general over-reaction to damage caused by the 1976 drought and many unnecessary repairs were undertaken. Underpinning of foundations is an extreme measure to adopt and may do more harm than good in some circumstances, for example, where swelling of the soil may take place afterwards through normal return of water to the soil in the wet season

Table 5.3 Visible damage through foundation failure

Degree of damage	Description
Very slight/slight	Fine cracks, not greater than 5 mm wide, often not visible in external brickwork and easily filled. Some slight sticking of doors and windows possible.
Moderate	Cracks may be typically from 5 to 15 mm wide; external brickwork will need repointing and some local replacement may be necessary. Doors and windows will stick and service pipes may fracture; general weather-tightness may be impaired.
Severe/very severe	Cracks will typically exceed 15 mm in width and may exceed 25 mm. Walls are likely to lean or bulge noticeably and may require shoring; beams may lose their bearing. Window frames and door frames will distort and glass is likely to break. Service pipes are likely to be disrupted. External repair work will be necessary, involving partial or complete rebuilding.

Based on 'Foundations for Low-Rise Buildings' (*The Structural Engineer*, June 1978).

or following the wholesale removal of trees and vegetation at the time when shrinkage damage first became apparent. Table 5.3 shows a classification of visible damage to walls and the related ease of repair of plaster and brickwork.

5.4 FILL

Good building land is scarce and this has put pressure on developers to fill other possible sites such as gravel pits, railway cuttings and open-cast mines. The support given by the fill depends crucially upon its type, the degree of consolidation it has reached and the way this has been achieved. All made-up ground should be treated as suspect because of the likelihood of extreme variability. The NHBC has recorded that the largest single cause of foundation failures to dwellings has been the use of poor fill [31]. Mostly difficulties have arisen through settlement of the fill following inadequate compaction.

5.4.1 Settlement of fill

A long time is usually needed for the natural settlement of fill, particularly if the predominant particle size is small. Slow consolidation occurs, too, when the fill has been inadequately broken or graded and contains excessive voids. Considerable compaction of originally loosely compacted fill can occur later if water reaches it, perhaps through a rise in the water-table level. Sites containing domestic refuse are especially hazardous, for these contain materials which may have large voids, such as old metal, glass and plastic containers, and also vegetable matter which eventually decomposes and results in subsidence of the super-imposed foundation. Shrinkable clay used as fill will shrink on drying and can cause settlement difficulties, particularly if construction takes place when the clay is saturated during wet weather.

5.4.2 Heaving of fill

While settlement of fill is the major cause of trouble when building is on made-up ground, swelling of shale used as fill has also caused extensive damage, though the problem seems not to be widespread in the UK. Swelling shales are known to have caused failures in the USA and in Canada. In this country, a series of failures occurred in the Teesside area and one in Glasgow in the 1970s. Investigations identified the ironstone shales which caused the trouble as belonging to the Whitbian shales of the Upper Lias, probably of the jet-rock series but also likely to be mixed with alum shales [34]. The cause of swelling was attributed to the oxidation of pyrites in the shale, resulting in a marked volume increase. The oxidation process also produces sulphuric acid and this reacts with calcite present in the shale to form gypsum. The crystallization of this gypsum between laminations in the rock is believed to be the predominant expansive force. The possibility of such problems occurring at points other than Teesside and Glasgow in the UK seems small but cannot be ruled out.

5.4.3 Effects of movement of fill

Movements due to settlement of fill are usually large and major cracking of external walls, screeds and internal partitions often results: doors and windows jam, gaps occur at heads of partitions and brickwork may bulge out. The expansive forces due to swelling pyritic shales cause concrete ground floors to lift and arch, and to crack. Internal walls and partitions lift and crack, and there may be outward movement of the perimeter walls near to DPC level.

5.4.4 Avoidance of damage by fill

Wherever possible, it is better to avoid sites which have been filled and all possible information about the site should be obtained by visits there, by discussion with local people and the local authority and by studies of local maps. Site visits should aim at observing signs of damage to any buildings bordering the site and should include the digging of trial pits to assess the nature of the soil. If it is clear that the site has been filled and cannot be avoided, then numerous trial pits will be needed to assess the nature and variability of the fill and its boundaries, its depth, its chemical composition, the degree of compaction, and the method by which the fill seems to have been laid and compacted. In addition, the level of the water table will need to be monitored for at least a year before building begins and the likelihood of further settlement should be assessed consequent upon the fill's becoming inundated. Adequate flexibility and protection to services to buildings will need to be provided. Specialist advice should be sought on ways of helping to consolidate the fill further and on the possible foundation solutions to the building to be erected by, for example, the use of piles.

Where chemical analysis has indicated the presence of pyrite and calcite, the best course to adopt is to remove the fill and replace it with a non-hazardous one. Where new fill is to be used on a site, the builder can, at least, exercise proper control over it. It should be of a granular nature, ideally a coarse sand or a gravel free from organic matter, and should be thoroughly compacted, layer by layer.

6

Floors, floor finishes and DPMs

Floors have not been a high risk area in recent years but failures have occurred, particularly in concrete screeds, by chemical attack, usually by sulphates, on the concrete base slab and through insufficient support to the slab by inadequately compacted hardcore.

6.1 HARDCORE

Hardcore is used to fill small depressions on sites and to adjust the amount of concrete needed in an over-site slab, following removal of topsoil from the site. It is also used on soft and wet sites to provide a good working surface and one which will not affect adversely the over-site concrete during placing. It has, too, some value in reducing moisture uptake from the ground. Hardcore is deemed to satisfy the Building Regulations (Section C4a) when it consists of clean clinker, broken brick or similar inert material free from water-soluble sulphates or other deleterious matter which might cause damage to the concrete. Materials mostly used in practice are concrete rubble, broken bricks and tiles, blast-furnace slag, various shales, pulverized fuel ash, quarry waste, chalk, gravel and crushed rock.

6.1.1 Hardcore and sulphate attack

A particular hazard to concrete floors can arise from the presence of soluble sulphates in hardcore or in the ground water. Solutions of sulphates can attack the set cement in concrete, as described in Chapter 3, the severity of attack much depending upon the type of sulphate present and the level of the water table. Broken bricks and tiles may contain soluble sulphates and, moreover, may be contaminated with gypsum plaster. Coal mining waste, too, usually contains soluble

Fig. 6.1 Sulphate attack on concrete floor.

sulphates. Burnt colliery waste from old tips tends to have a higher soluble sulphate content than unburnt spoil and, used as hardcore on wet sites, has frequently caused failure of over-site concrete. Other materials used as hardcore which can contain soluble sulphates include spent shales left as residue following the extraction of oil from oil shales, pulverized fuel ash, particularly if mixed with furnace bottom ash, blast-furnace slags derived from iron-making and shales containing pyrites (see Section 5.4.2).

Attack by sulphates on concrete floors, whether they derive from hardcore or from the soil, may be manifested initially by lifting of the floor and some binding of doors. As attack proceeds, major lifting and arching can occur, the concrete surface cracks and there may be some movement of the external walls near DPC level. These expansive forces are usually slow to develop and movements may not become apparent for several years.

Such sulphate attack may be avoided or greatly reduced in severity by using, for hardcore, materials such as coarse sand, gravel, crushed rock, clean concrete rubble and quarry waste, which are usually free from soluble sulphates; by ensuring that the concrete placed is of low permeability; and, if there is still a possibility of attack, by using a type of cement low in tricalcium aluminate, such as sulphate-resisting Portland cement complying with British Standard 4027 [12]. An estimate should be made of the sulphate content of samples of ground water. BRE Digest 250 [35] gives guidance on the selection of the type of cement and quality of concrete required to resist attack by sulphate-bearing ground waters and soils. It also provides information on the method for sampling and analysis of ground waters. Where soluble sulphates are present in hardcore, a water barrier, such as polyethylene sheet at least 0.2 mm thick should be placed to separate the floor slab from the hardcore.

6.1.2 Hardcore and swelling

Similar damage can be caused to floors by the swelling of hardcore due to factors other than sulphate formation. Materials likely to swell from these causes are principally slags derived from steel-making for these may include unhydrated lime or magnesia, some colliery spoils containing clay and refractory bricks used in chimneys and furnaces. These materials should not be used as hardcore.

6.1.3 Hardcore and compaction

If hardcore is not thoroughly compacted it will consolidate after the building has been completed and a solid ground-floor slab will no longer

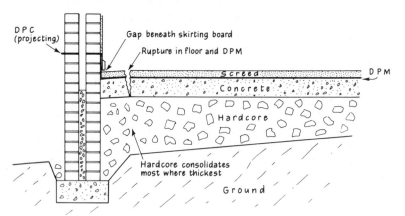

Fig. 6.2 Floor slabs may settle and crack if hardcore thickness is excessive.

be supported adequately over its whole area. This is particularly likely to happen when the depth of hardcore used is excessive as may occur on a sloping site, or where deep trench-fill has been used for foundations. Thorough compaction then becomes, if not impossible, at least unlikely. The solid slab is then likely to drop and crack particularly towards the edges under which the depth of hardcore is greatest. The first sign of trouble is usually the appearance of gaps between the floor and the skirting board. Cracks may appear in partitions. To avoid trouble, hardcore should be as well compacted as possible and a suspended floor, not a solid slab, should be used if the thickness of hardcore anywhere exceeds 600 mm.

6.2 DAMP-PROOFING OF FLOORS

To prevent the rising of moisture from the ground and into the floor finish, it is necessary to provide damp-proofing both in the walls and in solid ground floors. A horizontal DPC in walls is required to be not less than 150 mm above ground level. Concrete bases and screeds will allow ground moisture to pass through and it is necessary to provide a damp-proof membrane (DPM) either as a sandwich layer within the thickness of the concrete or upon the surface of the concrete. (In the latter case, the damp-proof layer usually provides the final floor finish, for example, pitch mastic or mastic asphalt.) These two requirements seem to be generally understood, though not always properly specified and achieved. A crucial need, however, is to link the DPC in the wall with that in the floor by means of a projection of the DPM (Fig. 6.3). This is often not done, possibly because the designer intended that this linkage should be provided but did not communicate the intention to the builder, feeling it to be a matter of normal good practice. Sometimes linkage has been made impossible by the power-float, used on the slab cast above the DPM, cutting off the projecting DPM. Whatever the reason, damp penetration has occurred in this manner in many post-war houses, a common effect being to cause decay of timber skirting boards.

Many materials may be used to provide damp-proofing for floors and BS CP 102 [36] provides information on their limitations. It also gives an indication of the properties of common flooring materials in relation to resistance to ground-moisture penetration. It may, however, be worth making the special point that cold applications of bituminous solutions, coal-tar pitch/rubber emulsions and bitumen/rubber emulsions will only give an impervious membrane if an adequate thickness is built up throughout. This will mean at least 0.6 mm and this will need two or

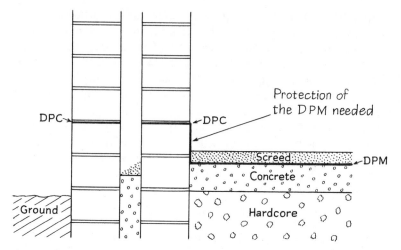

Fig. 6.3 Continuity between wall DPC and floor DPM. This is needed but often omitted.

more applications conscientiously brushed on – not easy to check after the event. Such cold applications must not be diluted. If polyethylene sheet is to be the DPM then it must have a minimum nominal thickness of 250 μm (0.25 mm) and a minimum spot thickness at any point of 200 μm and the laps between sheets must be double-welted. Special care and detailing will be needed to ensure the continuity of any DPM if this is punctured by services.

6.3 CONCRETE FLOORS

The design and construction of concrete floors, including granolithic and terrazzo, is dealt with in BS CP 204 [37]. The main problem with concrete floors, apart from sulphate attack, has occurred through failure to remember that concrete shrinks on drying and, in so doing, tends to crack and to exert stress at the interface between it and floor screeds and finishes. There has also been a failure to recognize that concrete does not present a surface which allows ready bonding of superimposed finishes.

6.3.1 Screeds

Most constructional problems have occurred with concrete floor screeds. It may well prove possible, however, to make the surface of a

Fig. 6.4 Curling and fracture of plastic tiles.

concrete base sufficiently level and smooth to accept the final floor finish without using a screed and this has much to recommend it. If a screed has to be applied, care is needed to get a good bond between it and the concrete base. If bond is poor, and shrinkage stresses are high, the screed will crack, with a tendency to curl at the edges of the cracks: if the screed is tapped the floor sounds hollow. Finishes applied over the concrete screed, such as tiles and sheet coverings, are also likely to crack and split and, ultimately, lose their adhesion to the screed. There are several factors which enhance the shrinkage and cracking of concrete screeds, and weaken its bond to the base concrete. The principal ones are inadequacies in the mix design of the screed and of the base on which it is to be cast; poor texture of the surface of the base concrete; too long an interval between casting the base concrete and the screed; an inadequate curing regime for the screed; too large an area of screed laid in one operation; and too thick a screed.

The mix proportions of the screed should be such that the ratio of cement to aggregate should not be greater than 1:3 nor less than 1:4.5 by weight of the dry materials. Suitable aggregates and the British Standards to which they should conform are listed in BS CP 204. The amount of water used should be just sufficient to allow the screed to be properly compacted. As water content increases so does screed shrinkage on drying and the amount of laitance which, when dry, shows as excessive dusting and crazing of the surface. The base concrete should have mix proportions not leaner than 1:2:4 of cement:dry fine aggregate:dry coarse aggregate by weight and should be thoroughly compacted to provide a firm and strong foundation.

A common cause of failure is a poor condition of the surface of the concrete base, leading to inadequate bonding of the screed. The texture required for the surface of the base depends partly upon the time interval between casting the base and the screed. If the screed is to be placed before the concrete base has set, all that is needed is for the surface of the base to be swept to remove cement laitance, water and any other material, such as leaves, which may lie upon it. Complete bonding should then be successfully obtained by this form of monolithic construction if the screed is laid within three hours of casting the base. Even though the screed and the base have different shrinkage characteristics, both can shrink together, which greatly reduces the tendency of the screed to curl and crack. The likelihood of the shrinkage forces exerted by the screed causing problems can be further minimized by ensuring that its thickness does not exceed 25 mm. It should, however, not be less than 10 mm thick. If a screed has to be placed, the monolithic form of construction is likely to give the greatest chance of success and should be used where possible and, particularly, if aggregates are used which are themselves liable to shrinkage (see Chapter 3). When monolithic construction is not possible, and the screed needs to be applied on a set concrete base, more rigorous treatment of the surface of the latter becomes necessary. This should be brushed with a stiff broom before it finally hardens, to remove laitance and loose aggregate, and to roughen the surface. Older concrete bases will need hacking and cleaning. The base concrete should be well wetted to reduce suction, preferably overnight, and a grout of cement and water brushed on, keeping just ahead of the application of the screed. A minimum screed thickness of 40 mm is needed. Where the screed is to be placed over a DPM then there can be no bonding with the base and the minimum screed thickness will need to be 50 mm at all points. A mechanical compactor or a heavy tamper must be used to obtain satisfactory compaction of the screed. The ordinary hand float will not give the necessary consolidation no matter how hard it is wielded. Screeds of thickness greater than 25 mm should be compacted in two layers one immediately after the other.

Floating screeds are those laid over a compressible layer of thermal or sound-insulating material. They are particularly prone to crushing and to cracking at gaps left between the individual slabs comprising the compressible layer. At such gaps, the screed can enter and form a solid bridge with the concrete base. To avoid crushing, floating screeds should be at least 65 mm thick or, if it contains heating elements, at least 75 mm thick. Penetration of the screed into joints can be prevented by laying impervious sheeting such as polyethylene to form a continuous layer over the compressible material. It is desirable to provide 20–50 mm mesh wire netting laid directly onto the sheeting to protect it, and the thermal or sound-insulating material, from mechanical damage when the screed is placed. To ensure that the compressible layer remains fully effective for its purpose, it should be turned up at the perimeter of the floor.

The screed generally needs to be trowelled to provide the smooth, dense surface required to take thin floor finishes, such as PVC tiles. Excessive trowelling can cause the surface to craze and, when dry, to dust. The effort required to produce this defect, however, suggests that this is one type of failure likely to decline with the passing years.

If large areas of screed are placed in one operation, random cracking often occurs. This can be reduced by laying screeds in separate bays and

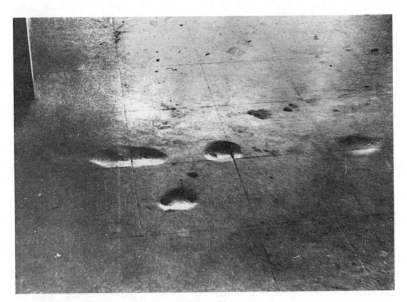

Fig. 6.5 Floating screed may crush if too thin.

was, indeed, the recommendation in the past. However, it proved difficult to prevent curling and slight unevenness at the junctions of bays and the resulting waviness showed through thin floor finishes. Not only did such undulations spoil the appearance of large areas of flooring but cracking of the thin finishes occurred. Remedial work required the removal of large areas of floor finish around the bay junction and the grinding down of the uneven edges, a costly procedure. It is better not to lay screeds in bays when thin floor finishes are to be applied. Dense screeds containing heating elements need special care to reduce the effects of shrinkage and laying in bays becomes necessary and specialist advice will need to be sought.

If concrete screeds dry rapidly, the risk and extent of cracking are increased, for the concrete is likely to lack strength sufficient to withstand the shrinkage forces. A simple protection, by covering with polyethylene sheeting or similar material for at least seven days, and then allowing the screed to dry naturally, is all that is needed but this is frequently neglected. The operation of screed laying and finishing is one where conscientious workmanship on site is vital for success but this is often lacking and has been the cause of many failures.

If poor adhesion between screed and base is suspected, and confirmed by the hollow sound generated when the former is tapped with a hammer, it does not follow that costly remedial measures are necessary. If hollowness is fairly local, it may not matter. Repair may only be necessary if the hollowness is accompanied by visible lifting of the screed, such that it is likely to break under the superimposed loads it is designed to take.

6.3.2 Granolithic concrete and terrazzo

Granolithic concrete is concrete suitable for use as a wearing surface and is made with aggregates specially selected to provide the surface hardness and texture required. It is used principally for factory floors and does not have any further finishing material laid upon it. The difficulties that have occurred with granolithic concrete are similar to those with screeds, namely, shrinkage cracking, poor bond to the base concrete and dusting of the surface. These difficulties are minimized in the ways already outlined in Section 6.3.1, with, once again, monolithic construction offering the best chance of success. To reduce curling to an acceptable amount, and because no thin floor finishes are to be applied, BS CP 204 recommends construction in bays, the maximum sizes of which depend upon the thickness of the concrete base and whether the granolithic concrete is to be laid in one monolithic operation. If it is, then maximum bay sizes should not exceed 30 m² where the thickness of the base concrete is at least 150 mm, or 15 m² for a 100 mm thick base. This smaller bay size is also recommended when the granolithic finish has to be laid separately on a set and hardened base. For success, granolithic concrete requires a high level of skill and close supervision and, unless these can be guaranteed, it is better avoided.

Terrazzo consists of a mix of cement and a decorative aggregate, usually marble, of a minimum size of 3 mm, and this is laid on a concrete screed. The failures that have occurred are of cracking and crazing, and lack of good bond to the screed. The risk of surface crazing is increased by too rapid drying. This has been a common problem caused by failure to cover the terrazzo with sheeting and by too great an absorption of water from the terrazzo mix into the screed. Surface crazing can be caused, also, by too rich a mix but the use of an excessive amount of cement seems hardly likely to be a common problem today.

Crazing and cracking may mar the decorative appearance of both granolithic concrete and terrazzo but are usually not otherwise serious. They are better tolerated than repaired, for this, particularly for terrazzo, is likely to be an expensive and specialized operation.

6.3.3 Shrinkage of suspended reinforced concrete floors

Suspended concrete floors usually deflect slightly and cause little trouble in so doing. Occasionally, greater deflection can occur, leading to cracking at the base of partitions because these are not then properly supported by the floor. This may, of course, be due to inadequate structural design or contruction but the possibility that the cause is the use of shrinkable aggregates (see Chapter 3) should also be considered, especially if such a failure occurs in Scotland.

6.3.4 Other forms of decay of concrete floors

Concrete is resistant to attack by the types and amounts of chemical likely to be used in a domestic environment but attack by acids, vegetable oils, fats, milk and sugar solutions can occur in the related industrial environments: mineral oils and greases are not troublesome. Whether or not attack is likely much depends upon the specific factory operations and each case must be assessed separately. There is a need, however, to protect the structural concrete by the use of an impermeable membrane, such as asphalt, bituminous felt or plastic sheeting, and to provide adequate drainage to the floor to facilitate rapid removal of spilled material.

6.4 MAGNESITE FLOORING

A floor finish not commonly used, but not yet extinct, consists of a mixture of calcined magnesite, together with a range of organic and inorganic fillers and pigments, gauged with a solution of magnesium chloride. This brew goes by the name of magnesium oxychloride or magnesite. The biggest problem arises because magnesium oxychloride is particularly susceptible to moisture and deteriorates when exposed to damp conditions. These conditions may arise if water penetrates the finish from below, perhaps, because of the omission of a DPM or the use of an inadequate one or, from above, from cleaning water or through condensation. When moisture attacks magnesium oxychloride, the surface and, indeed, the main body of the finish, disintegrates. Magnesium oxychloride floors have an inherent tendency to sweat, since magnesium chloride takes up moisture from the air. Beads of moisture appear on the surface, much as would occur through condensation, and sweating can be confused with the latter. Metalwork, for example, plumbing and electrical services, in contact with magnesium oxychloride is liable to corrode even when the floor is dry.

The durability of magnesium oxychloride flooring and its resistance to cracking and sweating are very dependent upon the mix proportions and upon the construction technique used. A common error is to use excessive amounts of the magnesium chloride solution with which to gauge the dry ingredients; another is failure to allow the floor to harden undisturbed for at least three days. Failure to maintain the surface by a good wax polish also enhances the risk of sweating, and excessive use of cleaning water, particularly if it is used with strong alkalis such as soda, will lead to gradual disintegation. This flooring needs very careful design and conscientious workmanship to perform successfully and guidance is given in BS CP 204 which, *inter alia*, requires that metalwork should be isolated from the floor finish at all times by at least 25 mm of uncracked dense concrete or by a coating of bitumen or coal-tar composition.

Once a magnesium oxychloride floor has started to disintegrate, it will be necessary, in general, to replace it. Sweating may possibly be overcome by improving ventilation, together with repeated applications of a good wax polish.

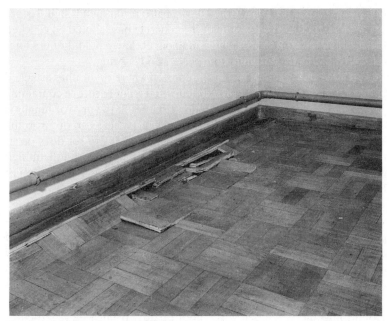

Fig. 6.6 Arching of wood blocks.

6.5 TIMBER FLOORING

Recommendations for laying timber flooring are contained in BS CP 201 [38]. Moisture poses the main problem, for this can cause decay and movement. The causes and effects of dry rot and wet rot are dealt with in Chapter 2 and general ways of minimizing the amounts of moisture present in Chapter 4. Specifically, timber floors, whether of board, strip, block or mosaic, need to be kept from becoming damp. Timber in contact with a concrete base needs to be protected by a damp-proof layer, either a surface or a sandwich membrane. Suitable materials are listed in BS CP 102. Failures have occurred not so much through omission of such membranes but more because of inadequate jointing with the wall DPC. The damp-proof membrane in the floor needs to be continuous with the DPC in the wall. When a sandwich membrane is used, it is necessary to allow sufficient time for the superimposed screed to dry before laying the timber finish upon it. The length of time required depends much upon individual circumstances but BS CP 201 gives rough guidance, that one should allow one month for every 25 mm of thickness of screed above the DPM. It also recommends that, before laying the wood floor, the state of the concrete base should be checked by a hygrometer or other reliable method. Relative humidity readings up to 80% indicate suitability to receive wood flooring.

In recent years, cases have been reported where dry rot has attacked timber floors and skirtings through infection from hardcore containing pieces of decayed timber. As stated in Chapter 2, the dry-rot fungus is adept at spreading and penetrating cracks, and these can occur in the site concrete when hardcore settles under load. If dry rot does occur through infection from timber in hardcore, eradication is difficult and costly. Prevention is greatly to be preferred, by thorough inspection of hardcore and the removal of any wood in it before use.

Suspended timber ground floors need to be adequately cross-ventilated by providing air-bricks to give at least 3000 mm^2 of open area per metre run of external wall. Building owners and, in particular, keen gardeners, should be careful not to obstruct the air-bricks in any way such as, for example, by the construction of an external patio. Good practice requires the provision of over-site concrete or other damp-resisting material to restrict the passage of water vapour from the ground into the space below the floor boards. It must be regarded as poor practice merely to remove topsoil from within the perimeter walls. DPCs are, of course, required in all sleeper walls.

The movement of timber as it absorbs and loses moisture is described in Chapter 3. A common defect with wood-block flooring is due to expansion of the blocks through uptake of moisture. When the forces associated with this expansion can no longer be accommodated by compression of the blocks, the latter will usually lift and arch, either separately from the adhesive which fixed them to the base concrete or by pulling the adhesive away from the concrete. The position of lifting may occur anywhere but is most common at the perimeter of the floor. Probably, the most common cause of moisture penetration in wood blocks is through failure to provide an adequate DPM under the blocks: too frequently, the adhesive used to stick them is taken, misguidedly, as providing an adequate barrier. Hot-applied bitumen or pitch used as the adhesive will give a proper barrier but single applications of cold bituminous emulsions will not. It is important, in order to minimize potential movement, that wood blocks and, indeed, wood flooring of all types, are laid at a moisture content likely to be close to that encountered in service. It is also necessary to provide a compression joint, for example, of cork, around the perimeter of the floor. If arching has occurred through dampness of a temporary nature, most commonly when the moisture in the blocks comes from construction water in the concrete screed, it will usually be possible to re-lay the blocks, when they, and the screed, have been given time to dry. The blocks should, of course, be examined for any signs of fungal attack before the decision is made to re-lay them. A less common defect is that caused by using blocks at too high an initial moisture content. When shrinkage occurs, gaps develop between blocks which spoil the appearance of the finished floor: there may also be some curling.

Joist hangers are commonly used to support timber joists but are often inadequately fixed principally by not bearing directly onto level masonry and by the back of the hanger not being tight against the face of the masonry. The joists used to support the overlying suspended timber floor may also be too short to sit properly within the hanger. In such cases the masonry into which the hanger has been built can be over-stressed leading to local crushing, particularly where low-strength

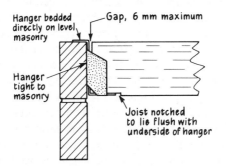

Fig. 6.7 Joist hanger and suspended timber floor: good practice giving firm and level support.

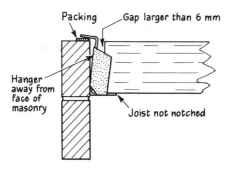

Fig. 6.8 Joist hanger and suspended timber floor: bad practice leading to local crushing of masonry and uneven floors.

blockwork has been used for the inner leaf. Additionally, the hanger may move under load and the floor can settle, becoming uneven and springy. Plaster around the perimeter of the ceiling may also crack. It is important that hangers are bedded directly onto level masonry without any packing, that the backs of hangers are tight to the inner face of the blockwork and that the gap between the end of the joist and the back plate of the hanger does not exceed 6 mm. Hangers used should match joist sizes and must be of a grade consistent with the masonry. They should never be used on masonry other than that for which they are designed. Joists should be notched out to sit flush with the underside of the hanger so that plasterboard for the ceiling can be fixed properly without distortion.

6.6 CLAY FLOOR TILES

Clay tiles suitable as floor finishes are described in BS 6431 [39]. The failures which have occurred have been caused by differential movement between the base concrete or screed and the tile finish. Burnt clay and concrete have somewhat different coefficients of thermal expansion (see Table 2.2) which lead to different amounts of movement with temperature changes. Freshly laid concrete bases or screeds shrink appreciably as they dry and this drying shrinkage may well be opposed to moisture expansion of the clay tiles, which occurs as moisture is absorbed by the tile. The moisture may be derived from the air, from cleaning water, from the ground if there are defects in damp-proofing (Section 6.2) and from the screed itself. This irreversible expansion of burnt-clay products is described in Chapter 3 and, while most of it occurs early in the life of the tile, later residual movements, if unable to be accommodated, can cause stresses to build up. Failures due to contraction of the concrete and expansion of the tile are manifested by arching of the tiles, often over a large area, or the forming of ridges, usually over one or two rows. In so doing, the tiles separate comparatively cleanly from the bedding and there is no general disintegration of either the tiles or the screed. Clearly, the best way to avoid failure is to minimize the movements which occur in the concrete screed or base and the tile, and to prevent direct transference of the stresses from one to the other. The worst conditions would occur if newly fired tiles were firmly bonded to a new concrete screed, for maximum moisture expansion and drying contraction movements could be expected. Tiles should not be used fresh from the kiln – even ageing of a fortnight will greatly reduce the irreversible moisture expansion.

Direct bonding between tile and base should be prevented and the best way of doing this is described in BS CP 202 [40]. The tiles are bedded in mortar onto a separating layer such as building paper, bitumen felt or polyethylene: sand should not be used as such a layer. These separating layers have the additional advantage of giving some resistance to the passage of moisture from the base into the tile. However, even with their use, movement joints are needed around the perimeter of the tiled area. Intermediate joints are needed in large areas and at points where stresses in the concrete base are most likely to occur, for example, over walls or beams in the case of tiled suspended floors. The concrete base and screed should be fully mature before tile laying begins and the floor should be closed to all traffic for four days after completion of tiling and then only foot traffic permitted for a further ten days.

6.7 PLASTIC SHEETS AND TILES

Sheet and tile flooring made from thermoplastic binders (principally PVC) are applied mostly to concrete screeds and the commonest cause of failure is through moisture passing from the screed. Such water contains alkalis derived from the concrete and these attack the adhesives used to stick the sheet or tile flooring. The adhesive becomes detached from the concrete surface though often remains well attached to the plastics flooring. The sheet or tile becomes loosened, the edges lift and damage may then be caused to them by normal traffic. Degraded adhesive, oozing through the joints, may also cause staining. Sodium carbonate can be left, following evaporation of water which has penetrated the joints in thermoplastic and vinyl asbestos tiles and, commonly, forms white bands around the edges of the tiles. Such a deposit will not, in many cases, prove harmful though it mars appearance. However, under particularly damp conditions, the salts may be absorbed into the body of the tiles and, on crystallizing, cause degradation of the edges.

The main reason for attack is through leaving insufficient time for the screed to dry before applying the sheet or tile and, at least, can be remedied easily. If, however, the DPM in the concrete floor is ineffective or absent, rising damp will continue to act adversely until the fault is remedied. With plastics sheet flooring, rising dampness through a defective DPM can also cause areas to become loose and to blister or ripple. A defective junction between the DPC in the wall and the DPM in the floor is likely to be the trouble if failures are localized near external walls, though this is not invariably so, for failures can occur some distance away from the entry point of the moisture. Excessive use of cleaning water can also cause loss of adhesion and some blistering.

Sometimes, inadequate bond between adhesive and tile can occur even when conditions are dry. This is caused by too long a delay between spreading the adhesive and laying the tile, leading to loss of surface tack. This is a fault of workmanship and is manifested principally by the looseness of individual tiles and an absence of staining. Some adhesives can cause flexible PVC flooring to shrink through migration of plasticizer into the adhesive, the latter being softened as a consequence. Under traffic, large gaps can then appear between tiles. This can happen, too, if solvent contained in the adhesive has become entrapped, through too early an application of large areas of sheeting.

Most failures with plastics sheet and tile flooring can be prevented if care is taken to ensure that there is an effective DPM; that the screed is allowed to dry properly before application; that the use of cleaning water is kept to a minimum; that the finish is applied neither too soon nor too long after application of the adhesive; and that only those adhesives specifically recommended by the manufacturer of the tile or sheet finish are used.

The repair which will be feasible should failures happen will much depend upon the cause. It may be possible to re-use loosened tiles or sheeting if all that is needed is to give more time for the screed to dry. The absence or defectiveness of a DPM, however, may involve more radical repair which, in turn, may necessitate a change in the type of floor finish used.

7

Walls and DPCs

This chapter deals with problems affecting walls and their DPCs but leaves consideration of claddings on walls and of failures associated with openings at windows and doors to the succeeding two chapters. Renderings are, however, more conveniently dealt with here.

Failures are essentially of two types – failure to provide adequate protection against moisture penetration, and cracking or spalling of the walling materials.

7.1 MOISTURE PENETRATION FROM THE GROUND

Rising dampness in walls is likely to cause damage to internal plaster and decorations, particularly when hygroscopic ground salts are brought up in solution, as they generally are. The line of dampness due to the complete absence of a DPC is usually fairly continuous and roughly horizontal, and can extend several feet in bad cases. Remedial measures much depend upon the nature of the wall. If they are of brickwork or coursed stonework, it may be feasible to cut a slot in the wall and insert a DPC or to inject a chemical damp-proofing system. Old walls of random masonry construction are unlikely to yield to these treatments and these can be nearly impossible to damp-proof. However, this book is concerned principally with the avoidance of failure in new construction and one would not expect the complete absence of a DPC in such work. A more common problem in modern construction is bridging of the cavity wall by excessive mortar droppings at the bottom of the wall (see Fig. 7.1). This leads to irregular damp patches appearing adjacent to the mortar droppings. If such poor workmanship cannot be prevented then, fortunately, the remedy is not too expensive. Bricks will need to be cut away and the cavity cleaned out.

Inadequate laps in flexible DPCs can also lead to moisture penetration. Laps should be of at least 100 mm and the DPC laid on a full mortar bed and also fully covered with mortar above to prevent damage. It is

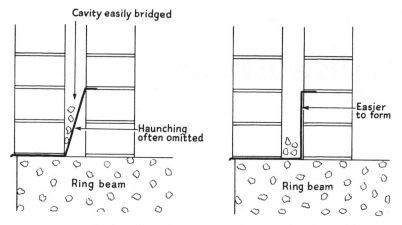

Fig. 7.1 Bridging of wall DPC. DPC is seldom supported; mortar droppings are difficult to remove.

Fig. 7.2 Preferred wall DPC. DPC needs no haunching; more droppings are necessary before DPC is bridged.

not uncommon for the DPC in the external wall to have been bridged by pointing, or rendering, over it. This should be prevented by projecting the DPC slightly through the external face and by stopping any rendering short of it. Bridging can also occur through over-zealous gardening, a common cause being the heaping of soil against the wall or the construction of patio paving above DPC level. Such gardening activities, if afterwards remedied, will give the home owner two lots of backache and a stained wall.

7.2 RAIN PENETRATION

Eaves, canopies, cornices, string courses and other forms of overhang have been used traditionally to protect walls from rain and its effects but in post-war years their use has declined. Flat roofs have become more frequent and, where pitched roofs still prevail, eaves overhangs have diminished. The extent to which such features affect the degree of wetness of walls, and the time for which they remain wet, has not been adequately researched. Though overhangs can affect the air stream, particularly in tall buildings, with uncertain effects, it is probable that walls in general are wetter for longer when they are absent.

The ability of a wall to exclude rain thus depends partly upon such design features but principally upon its topographical and geographical situation. Table 10 of BS 5628 Part 3 [41] gives exposure categories either in terms of the local spell indices calculated using BS DD 93 [24] or those based on the BRE driving rain index [23]. It also depends greatly upon the materials and construction of the wall. Generally speaking, rain is likely to penetrate solid brickwork or blockwork which is not rendered, usually directly through cracks in the mortar, and between mortar and brick or block. The risk is increased if the mortar is of poor quality and if the vertical joints of masonry are filled inadequately. Rain penetration shows as damp patches on the internal face of the wall usually within a few hours of rain falling: when the wall dries, a stain is often left. Solid masonry walls are now seldom built and little more need be stated here, except to warn against some possible difficulties if water-repellent solutions are applied to reduce the risk of rain penetration in existing walls. The trouble with such coatings is that they retard considerably the subsequent evaporation of any water that may get in through lack of continuity in the coating or through unfilled cracks and, in so doing, can lead to more persistent dampness. To be successful, a continuous, crack-free coating is needed, which may be difficult to achieve. When soluble salts are present in a wall, they can, when the wall remains untreated, move in solution to the outer face and, under dry conditions, appear on the surface as an efflorescence. A water repellent may prevent this free movement, and salts can be deposited within the pores of the masonry and some distance beneath the surface. When they crystallize on drying, the forces produced can cause spalling of the surface. A preferred treatment, though admittedly one which changes the appearance of the wall, is to clad it with a porous rendering or with tiles.

Cavity walls are now used for the great majority of buildings and should, in theory, prevent the direct penetration of rain to the inner leaf. In general, they do, but plenty of problems have occurred through bridging of the cavity by mortar droppings; by poor positioning of wall ties; by unwise use of cavity fill for thermal insulation; and by inadequate design and construction of DPCs at junctions. It is worth making the point that rain will penetrate the outer leaf of most walls and run down the inner face of that leaf. Penetration is likely to develop quickly and to an appreciable extent in some types of wall, particularly concrete masonry block walls.

BS 5628 Part 3 calls for any mortar which unavoidably falls on the wall ties and cavity trays to be removed daily. This is not a counsel of perfection but a statement of what used to be a normal brick-laying operation. If it is not done, damp patches are likely to show on the inner leaf within a few hours of rainfall. Such patches are likely to occur sporadically and the use of a metal detector will, most probably, indicate that a tie is close by. Remedial measures usually involve taking bricks from the outer leaf just above the site of the dampness and removing the mortar, or even sometimes broken bricks, which have been dropped down the cavity.

Ties should not be bedded with a fall from the outer to the inner leaf but sometimes are, and, in these cases, can allow rain water to trickle down the tie to the inner leaf. The fault can be due to failure to match courses by bringing up one leaf of the wall too far in height above the other during construction. Remedial measures will necessitate the re-bedding of the ties but whether this will prove possible will much depend upon the degree of mismatch of the mortar joints. Ties are, not infrequently, poorly positioned laterally so that the central projection in the tie, designed to shed any water penetrating the outer leaf, is not in the centre of the cavity but close to, or touching, the inner leaf. Fortunately these faults, or at least the manifestation of them, are not as common as the lodgement of mortar droppings, for they are inherently more difficult and expensive to put right.

7.2.1 Cavity fill for existing buildings

In recent years, the need to conserve energy has led to the provision of thermal insulation within the cavity of a cavity wall. For existing buildings a commonly used material is urea-formaldehyde, foamed on site and injected into the cavity to fill it. As the main purpose of the cavity is to prevent rain penetration, the filling of it clearly, in principle, increases the risk. Whether or not rain penetration to the inner leaf will occur depends upon many factors, of which the main one is the adequacy of the external wall against wind-driven rain. Rain can pass through fissures in the foam. These can arise when urea-formaldehyde sets and shrinks. The extent to which such fissuring can occur depends principally upon the degree of control exercised during injection and upon the precise formulation of the foam. Further risks of penetration arise if the upper boundary of the foam is not fully protected from the ingress of rain.

There have been failures due to the operation of one or more of these factors, though numbers so far have not been great. When they occur to the extent which leads to visible signs on the inner leaf, these take the form of patches of damp associated with periods of rainfall. They cannot readily be distinguished from rain penetration caused by dirty wall ties.

The risk will be minimized if, before filling the cavity, the suitability of the materials and construction of the external leaf is first checked against the local driving-rain index value. BS 5618 [42], which is a code of practice for the use of urea-formaldehyde foam as a cavity fill, provides such design guidance and it is essential that the criteria stated in that Standard are followed. BS 5618 should be read in conjunction with BS 8208 Part 1 [43] which covers those aspects which should be taken into account when assessing the general suitability of cavity walls for insulation. In some areas with a high driving-rain index, it may prove necessary to up-grade the external walls, for example, by rendering.

There is, as yet, little experience of the use of expanded polystyrene pellets and polyurethane foams, which may be used to fill cavities. Prospective users of such materials would do well to use only those which have a British Board of Agrement (BBA) certificate and to check that the conditions of exposure to be encountered will be no worse than those to which the certificate relates.

When rain penetration does occur with urea-formaldehyde foam, it may be slight and, if so, it may be possible to remedy it locally by refilling the cavity at the appropriate spot. Otherwise, it will probably be necessary to increase the general resistance of the outer leaf to rain penetration by a suitable rendering or cladding. The expense of these latter remedies, however, may outweigh the financial advantages obtained by insulating: cavity-filling is an operation which it pays to get right first time.

Mineral rock fibre treated with water repellent has also been used in regions with a high driving-rain index and, if properly injected, does not allow rain to penetrate. With all types of cavity fill it is essential that the outer skin of the structure is in sound condition otherwise spalling of the outer skin may occur after filling. It is also essential that cavities are filled by experienced and specialist tradesmen to minimize the risk of cavities becoming blocked by rubble falling from the face of the wall being drilled.

7.2.2 Built-in cavity insulation

The more stringent present-day requirements for thermal insulation can often be met conveniently by partially filling the cavity walls of buildings under construction with insulation boards or by wholly filling them with insulation batts. As might be anticipated, fewer problems are likely to occur with partial filling as there is at least some cavity remaining. The boards mainly used are of expanded polystyrene but foamed plastics and glass fibre are also used. The boards are fixed to the cavity face of the inner leaf by clips attached to the wall ties or by nailing. If the number of clips or nails is inadequate the boards will not lie firmly and flatly against the wall and will project into the cavity. The smaller the designed width of cavity the greater will be the risk that the projection will be sufficient to channel any drips from above onto the inner leaf. Of course any such risks will be enhanced if the defects already mentioned in Section 7.2 are also present. When clips attached to the wall ties are used four will be needed for each board and wall tie spacings will need to match the dimensions of the boards. If boards are nailed, six nails per board will be the minimum needed. Again, prospective users would do well to use those boards which have a BBA certificate and to follow the restrictions which these certificates impose upon use. These are more stringent where the designed clear cavity is to be less than 50 mm. Then, if boards do not extend to the full height of the wall, cavity trays will be

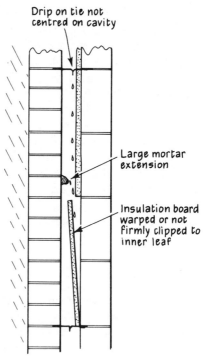

Fig. 7.3 Bad practice: tie not centred; mortar extruded; board warped or poorly fixed. Rain is fed to inner leaf.

needed with weep holes above (see Section 7.2.3) to prevent rain from possibly dripping onto the top edge of the board and thence being channelled into the inner leaf. Insulation boards cannot be expected to lie flat against the wall if they are warped and this will happen if they are stored over battens; they must be stored flat.

Insulation batts used for total cavity fill are made from layers of glass fibres or mineral fibres treated with a water repellent. They have a somewhat laminated structure. There is, as yet, no great body of experience of rain penetration associated with their use but recent research has pointed to the likely hazards [44]. The main risk is when mortar used in the brickwork of the outer leaf extrudes into the cavity and adjacent to a joint between the batts. Such an extrusion will compress the batt as it is pushed in and will deform the laminations towards the inner leaf. Water entering through the outer leaf may drain down between laminations in the batts and follow these deformations to

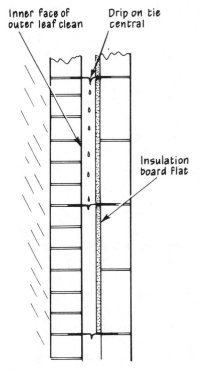

Fig. 7.4 Good practice: tie centred; inner face of outer leaf clean; board flat and properly clipped. No rain penetration.

become closer to, and possibly reach, the inner leaf. It will be necessary to ensure that any such extrusions of mortar at the inner face of the outer leaf, but particularly those adjacent to horizontal joints between batts, are cleaned off as bricklaying proceeds. This will always be easier if the external brick face is taken up in advance of the inner leaf. If insulation batts are not taken up to the full height of the wall, cavity trays must be used with weep holes to protect the top edge of the batts. The batts themselves are not very robust and must be stored carefully under cover and not damaged.

7.2.3 Cavity trays at junctions

There have been numerous cases of rain penetration reported at junctions between walls and windows and doors, and where walls adjoin solid floors and ring beams. The former problem is considered in Chapter 9, which deals with external joinery. It is, however, appropriate to note here some difficulties which have been encountered when walls form a junction with concrete ring beams, columns and floors. Where cavity walls are built off concrete ring beams or floors, rain will usually penetrate the outer leaf and run down, and a cavity tray is needed to prevent the water from collecting on the beam or floor and feeding through the inner leaf, or along the floor, to the inside of the building. Cavity trays are also needed where the external leaf of a building becomes part of an internal wall at a lower level, a common occurrence when extensions are added to houses and in the construction of stepped terrace blocks. They are also needed at other points where the cavity is bridged, for example above airbricks and meter boxes. Many failures have occurred at these junctions for a variety of reasons. Sometimes cavity trays have not been provided at all which is a sure recipe for disaster. It should be noted here that house extensions and stepped terrace construction may be such that there is no room for a cavity tray of adequate depth to be provided. In such cases it is essential that the portion of wall which remains external is protected from rain penetration by other means. This may be by the roof above overhanging sufficiently or by cladding the external portion by a rendering or tile hanging. Other common causes of rain penetration have been the failure to provide a sufficient depth of upstand to the cavity tray; failure to prevent bridging of the cavity by mortar droppings; damage to the flexible materials mostly used, nowadays, in the construction of the tray; failure to support the tray by a mortar bed (haunching); inadequate lapping or sealing of lengths of cavity tray; and omission of stop-ends.

The traditional detail, with the tray sloping across the cavity, is prone to assist blockage of the cavity and experience has indicated that a 75 mm upstand, previously assumed to be adequate, can be bridged by mortar droppings. Moreover, the traditional cavity tray was of metal and supported in its diagonal passage across the cavity by a mortar haunching. Nowadays, it may be based on bitumen or polyethylene and the haunching is frequently omitted or poorly placed, causing the flexible tray to sag through lack of support. If mortar droppings are removed, there is a considerable risk of puncturing or distorting the tray, particularly where it is unsupported, thus making it ineffective. Rain has penetrated when lengths of the material used to form the tray have merely been overlapped and not properly sealed. Efficient sealing is, of course, difficult when the material is not supported. Penetration underneath the cavity tray is also a risk if this is not properly bedded, particularly if the surface of the concrete ring beam or floor is rough. Where cavity walls bear against columns, rain can penetrate at the junction between the cavity tray and the column face in the absence of stop-ends sealed to the latter.

Penetration from these causes is manifested by damp patches appearing after rain at or near to floor level and columns. The position of the patches may, or may not, however, be close to that of the original entry of the rain.

Failures may be prevented by designing cavity trays carefully in relation to the details of the construction involved. It should not be left to the man on site to 'mock up' a solution: nowadays, this is not likely to result in success. Some general recommendations are that the upstand to the cavity tray should be a minimum of 150 mm; the tray should be dressed down the outer face of the inner leaf and then across the cavity – that is, in an 'L' shape for the tray rather than the traditional 'Z' shape shown in Fig. 7.1 (see Fig. 7.2); it should go over, not under, any flashings or DPCs associated with the detail; the tray should be bedded in wet mortar; lengths should be sealed one over the other, not just lapped; and stop-ends should be provided at junctions with columns. Duell and Lawson [45] provide useful illustrations and information on both the design and installation of cavity trays and other forms of DPC.

Water penetrating along the external leaf and collecting above a DPC at the base of the wall, or above a cavity tray, needs to be drained to the outside. Weepholes have been the traditional method of drainage, formed by omitting the mortar in appropriate perpend joints between bricks, usually every fourth joint directly above the DPC or cavity tray. Considerable amounts of water can collect in the absence of such weepholes and can feed through weak points in the system, particularly

Fig. 7.5 Cavity tray sagging and unsupported.

between unsealed laps in the DPC. Sometimes, weepholes have been provided but have been blocked by mortar droppings or cavity fill and made partially or wholly ineffective. If drainage cannot occur through weepholes, the water may drain out through the mortar joint above the DPC and it is probable that in many situations this can prove adequate. In exposed situations, however, some more positive drainage is needed. To prevent the possibility that rain will be blown into cavities through weepholes on very exposed sites, Duell and Lawson refer to the use of tubes bent to prevent blow-back. They also expose the need to understand more fully the value of weepholes, but believe that it is safest to provide some form of drainage to cavities. This certainly seems desirable.

The elimination of defects in the DPC system at the junctions mentioned is costly and most difficult, and greatly outweighs the extra effort needed initially to get design and construction right. Even the comparatively simple task of drilling out weepholes may cause damage to the cavity tray. It is recommended that pre-formed trays are used wherever possible and that all cavity trays are installed with stop-ends properly sealed to the tray.

7.3 COLD BRIDGES AND INTERSTITIAL CONDENSATION

Bridging of a cavity wall by dense materials like concrete can result in cold areas and lead to condensation upon them. Areas particularly at risk include the surfaces of concrete lintels, ring beams, and floor slabs which pass through to the external wall. The dampness will, in general, be more widespread in area than that caused by mortar droppings on ties, and will be localized at the level corresponding to the feature which bridges the cavity. It will, moreover, appear unassociated with rainfall. Such dampness needs to be avoided or minimized for the reasons given in Chapter 4.

It may be possible to avoid cold bridging by designing so that some cavity is kept, or by providing extra thermal insulation on the inner side of the bridging feature. Such preventive measures are more readily incorporated at the design stage than allowed for later, though it may be possible to protect dense lintels and ring beams by added thermal insulation.

The main points concerning interstitial condensation have already been made in Section 4.5. It only needs to be stated here that where vapour barriers are used they must be placed on the warm side of the insulation and that wherever possible the permeability to water vapour of the external walls of a structure should increase towards the outside. Special care will be needed to ensure the continuity of vapour barriers at discontinuities in the structure.

7.4 CRACKING AND SPALLING OF MASONRY THROUGH MOVEMENT

Load-bearing masonry walls can crack and their surfaces may spall. Although such defects have not been amongst the most numerous, nationally, in recent years, there have been serious localized problems. There are reasons for believing that current construction methods and materials may lead to an increase in sensitivity to some of the agencies which give rise to these defects. Cracking of masonry is associated with movement and a principal cause is changes in moisture content of the units.

7.4.1 Movement due to moisture changes in the masonry units

Reversible and irreversible moisture movements in fired clay bricks are described in Chapter 3. Reversible dimensional changes are shown to be small – of the order of 0.02% – and such changes do not cause problems in practice. It is the irreversible movement, which is usually many times as great, that has led to cracking. The extent of movement depends upon the type of clay, the degree of firing and the time which has elapsed

Fig. 7.6 Cracking caused by moisture expansion of bricks.

since the bricks were removed from the kiln. As soon as bricks are removed and cool, this expansion will start. Typical expansion in the first two days can range from 0.02% for bricks made from London and Gault clays to 0.08% for those made from Weald clays. Fletton bricks made from the lower Oxford clays have a typical two-day expansion of 0.03%. Such movement is around half the long-term, irreversible movement. As mentioned earlier, the corresponding expansion of brick walls as opposed to bricks is likely to be only one half of the latter. This should be taken as an approximate value for the type of mortar and the restraint imposed upon the wall by the form of construction can also have an effect.

This irreversible moisture movement may cause several visible defects in brickwork. One typical result is an over-sailing of bituminous felt and polyethylene DPCs at the end of the wall, for such materials provide little restraint to expansion. This lack of restraint may also lead to cracking of walls near quoins. The sections of brickwork on either side of the return wall expand and tend to rotate the return. If this is short, for example less than 600 mm, vertical cracks develop, typically straight, and visible above DPC only, but likely to extend the height of the building. Cracking from the same cause and of similar appearance can occur near the corner of a building.

The majority of such problems can be overcome by not building clay bricks into a wall until at least three days have elapsed after withdrawal from the kiln and by avoiding the use of short return walls in long runs of brickwork. It should be noted that hosing down stacks of bricks and dipping individual bricks into water before laying is ineffective in preventing subsequent expansion. BS 5628 Part 3 calls for runs of clay bricks in walling to have a joint capable of accommodating 10 mm of movement about every 12 metres. As a general guide the width of joint in mm should be about 30% more than the distance between joints in metres.

Where oversailing of the DPC in the length of the wall has occurred, there is little that can be done but the movement is unlikely to be such as to cause instability. Cracking near quoins is also unlikely to be of structural significance and, when the cracks are fine, no repair is necessary. Where cracks are wide, it will usually be desirable to replace the cracked bricks. As irreversible expansion will be over effectively by then, the defect will not recur.

Calcium silicate and concrete bricks and blocks shrink upon drying and curing after manufacture and are also subject to reversible moisture movement, for which allowance needs to be made if defects are to be avoided. These defects take the form of vertical or diagonal cracks which commonly run from the corners of openings in walls, for example, between upper and lower windows. If the mortar used is not too strong, the cracks will pass through the joints: otherwise, they may pass through the bricks or blocks, which is more disfiguring. Vertical cracks passing through bricks is a good indication that too strong a mortar has been used. Cracks appear early in the life of the building and their width is near to its maximum within a year.

Calcium silicate bricks and concrete blocks have a high water absorption and many of the problems are caused by leaving them on site unprotected from the rain. They should be transported and stored under cover, be at least two weeks old and, also, be laid as dry as practicable. The tops of unfinished walling should be protected from rain. A mortar generally not stronger than a 1:2:9 cement:lime:sand mix is recommended unless frost is likely during construction, when a 1:1:6 cement:lime:sand mix or its equivalent in characteristics may be necessary. For calcium silicate brickwork movement joints should be provided every 7.5 metres, though intervals of 9 metres may be satisfactory if the mortar has been chosen correctly. For concrete brickwork or blockwork a close spacing of movement joints to about 6 metres is preferable. Movement joints should be located at points in the structure where lateral support is provided. In long runs of terrace housing, where the problem most commonly occurs, these will be at separating walls between the houses. Extra ties should be put in at 300 mm vertical spacing at each side of the joint and the more flexible the wall tie the better, provided, of course, that the structural needs of the design are fully met.

Cracking as a result of movement is repaired by either repointing the mortar joints or by cutting out and replacing cracked bricks or blocks.

7.4.2 Movement due to temperature changes

Typical thermal movement of masonry walling is small, about 0.01% for a change in temperature of 20°C. Such movement does not cause defects in walls unless temperature changes are exceptional, therefore, it need not be considered further here.

7.4.3 Cracking due to roof movement

Cracking of walls caused by the spread of pitched roofs is not a common defect with new buildings. It occurs mainly through weakening of the structural members of a roof with time. Another cause, apart from general under-design, which is unlikely today, is the substitution of much heavier roof tiles for those originally specified. As the roof spreads, it moves the top few courses of the wall masonry outwards slightly; horizontal cracks appear in the plaster on the internal wall close to eaves level and cracking may also be noticeable externally. The main precaution necessary, nowadays, is to ensure that roof finishes are not changed without checking against the original design intentions. Should a defect from this cause appear, a detailed inspection of the state of the roof timbers, of the roof finish and the quality of the mortar joints will be needed before remedial work can be decided upon. It may well prove necessary to rebuild the roof and the top few courses of masonry.

A more common defect associated with roof movement is that caused by the movement of flat concrete roofs. These can move appreciably with diurnal changes in temperature, particularly if the surface finish is dark in colour. The roof then tends to move outwards at the top of the walls and may push the top courses of masonry out slightly. Cracking then occurs within the internal plaster finish, usually near to the corners of the building and close to the junction of the roof with the walls. Some cracking in the mortar joints of the external wall just beneath the roof may also be seen. It may be possible to prevent or minimize such defects by ensuring that any roof likely to exert a horizontal thrust on the wall is separated from it by a flexible DPC. When this is not feasible, the roof and the wall should be designed to act together. In either case, it will be advantageous to provide a solar reflective finish to the roof to minimize movement.

This defect is not of great significance, generally affecting only internal appearance. Where cracks are horizontal, and at the wall/roof junction, a coving may hide them. Decorating with lining paper and wallpaper may also prove successful to hide cracking which can occur lower down, probably near to the top of window openings.

7.5 DAMAGE TO WALLS BY CHEMICAL ATTACK

Most ordinary clay bricks contain sulphates of sodium, magnesium or calcium. These salts are soluble in water, calcium sulphate being less soluble than the other two. Normally, these sulphates are seen as the harmless efflorescence noted in Chapter 3, affect appearance only and need simply to be brushed away. Under wet conditions, however, the salts can react with tricalcium aluminate, which is always present in ordinary Portland cement, to form calcium sulphoaluminate. Ordinary

Portland cements vary in the amounts of tricalcium aluminate they contain, those with the highest amounts having the least resistance to attack by the soluble sulphates. The reaction is accompanied by a marked volume increase. Wet ordinary bricks, therefore, in contact with mortar based on ordinary Portland cement, may lead to the formation of calcium sulphoaluminate and to disruptive expansion. It is, in fact, the mortar which is attacked, not the bricks, and the volume increase consequent upon the formation of calcium sulphoaluminate can cause vertical expansion of brickwork, which is commonly as high as 0.2%. In theory, most brick walls with mortars based on ordinary Portland cement are liable to sulphate attack. In practice, fortunately, considerable sustained wetting is necessary. The most vulnerable walls are earth-retaining walls and parapet walls but sulphate attack is a problem also on rendered, and on facing, brickwork.

The main sign of trouble on facing brickwork is cracking in the horizontal mortar joints, which generally occurs in a number of them. This may be preceded by some horizontal cracking in the plaster of plastered internal leaves of cavity walls, usually near to eaves level. This internal cracking is caused by the expansion of the outer leaf in which the reaction is taking place, putting the inner leaf into tension. As the reaction proceeds, the external mortar joint spalls at the surface, and becomes weak and friable. The vertical expansion can result ultimately in spalling of the surface of the facing bricks and some bowing of the external walls. The brickwork may oversail the DPC at the corners of the building, in a manner similar to that caused by the irreversible moisture expansion commented upon in Section 7.4.1. However, the latter movement will take place in the early months of the life of the building, while sulphate expansion is unlikely to appear for several years.

On rendered brickwork, sulphate attack is manifested by cracking of the rendering, the cracks being mainly horizontal and corresponding to the mortar joints below. The rendering may adhere quite well to the bricks early in the attack but areas are likely to become detached as the expansion of the underlying brickwork causes severance of the bond between the two materials.

There are three main ways of preventing sulphate attack in mortars: by ensuring that walls do not get, and stay, unduly wetted; by selecting bricks low in soluble sulphates; and by the use of cements low in tricalcium aluminate.

In most parts of the UK, the walls of structures between DPC and eaves level can be prevented from getting unduly wet by following established good design and construction procedures. These include the need to provide a good overhang to eaves and verges, to ensure that DPCs, flashings and weathering systems are effective and by keeping brickwork and blockwork as dry as feasible before, and during, construction. In areas where the driving-rain index is high (see Chapter 4), special precautions will be needed. In such areas, it will be desirable to give additional protection to the walls by rendering, tile hanging or similar treatment.

Renderings generally keep the walls beneath them dry, and well-designed and applied renderings will do so and help prevent sulphate attack. Those liable to shrinkage cracking, however, and dense renderings are particularly liable, can allow rain to penetrate the cracks but not escape readily afterwards. Under such conditions, the brick-work can remain wet for long periods, which assists the disruptive reaction to take place. The rendering itself may be attacked: expansion

Fig. 7.7 Shrinkage cracking and sulphate attack on rendering.

can then exceed 0.2%. A 1:1:5 to 6 cement:lime:sand mix, or its equivalent, would be generally suitable for most conditions: denser, stronger mixes should not be used. Comprehensive recommendations for the design and execution of renderings on all common backgrounds are covered by BS 5262 [46].

It is possible to obtain bricks low in soluble sulphates by specifying those designated as of 'special quality' in BS 3921. Such a specification is likely to be considered too restrictive, however, for walling between DPC and eaves level but may well be of value for brickwork severely exposed to rain, such as in chimneys and parapets. This is considered in Chapter 10, which is concerned principally with roofs, but in the context of which problems with chimneys and parapets are more conveniently covered.

Sulphate-resisting Portland cement and supersulphated cements can be obtained and, if these are used instead of ordinary Portland cement in the mortar, the likelihood of sulphate attack can be greatly reduced.

Where unrendered brickwork has expanded through sulphate attack, remedial measures will much depend upon the extent of damage. A first essential is to prevent the brickwork from continuing to become wet (and to eliminate the source of dampness, if this is a defect such as a leaking gutter or a defective downpipe). Generally, and in the absence of such obvious causes of dampness, it will be necessary to resort to tile hanging, ship-lap boarding or similar external treatment. When mortar has deteriorated considerably, repointing using mortars of composition 1:0.5:4.5 cement:lime:sand or 1:5 to 6 cement:sand with a plasticizer will be desirable. Treatment with a surface waterproofer is unlikely to be effective. Such treatment may, indeed, promote spalling of bricks containing large amounts of sulphates. They will, moreover, need frequent renewal.

When the rendering on masonry has deteriorated through sulphate attack it, too, will continue to do so unless the source of dampness is eliminated and moisture is prevented from reaching the brickwork. When decay is localized, as might be the case if the cause were a leaking gutter, then some cutting-out and re-rendering may be feasible, once the cause has been dealt with adequately. Where sulphate attack is general, nothing will be gained by attempting to repair the rendering. It will be necessary to prevent rain from reaching the brickwork by tile hanging or some similar technique.

Brickwork is not affected by atmospheric pollution but stonework is, and the causes and defects are described in Chapter 3. When decay is superficial, it may be arrested by having the stonework cleaned by a specialist firm, which will use the most appropriate method in relation to the type of stone used, the extent of decay and the nature of the building. Such firms can also restore decayed surfaces by plastic repair, but serious decay will necessitate removal and replacement of the whole, or a substantial part, of the stone block. It is generally not feasible to arrest the decay of stonework by the use of a stone preservative. Only for small areas of stonework of historical or architectural significance may it be possible, depending upon the particular types of salt contaminating the stone.

7.6 DAMAGE TO WALLS BY PHYSICAL ATTACK

Water freezing within the pores of porous building materials, such as brick, stone, mortar and concrete, exerts a force, the magnitude of which much depends upon the amount of water present and the pore structure of the material. These expansive forces can result in defects but, as stated in Chapter 3, frost damage is comparatively rare between DPC and eaves level, though not uncommon in parapets. The winter of 1978–79, however, produced more failures than usual, which may have

Fig. 7.8 Frost damage to bricks.

been due to unusual combinations of sub-zero temperatures following prolonged rain. Under-fired bricks are more susceptible to frost damage than well-fired bricks and it is conceivable that over-enthusiastic attention to fuel economy may have contributed to the problem. In the general context of building failures, frost damage to walls has been of minor importance but the position needs to be watched with some care if winters become more severe, fuel economy remains a key issue and buildings become more highly insulated, for external walls may then reach lower temperatures more often and for longer periods.

Damage in buildings mainly occurs to brick, stone and mortar: concrete is seldom affected. Brick damage is seen usually as a crumbling or flaking of the surface and mortar suffers similar damage, which usually occurs before the mortar has had time to harden. Hardened mortar, though not immune if of incorrect strength, is not often attacked. Frost damage on limestone mostly causes small pieces to spall away but major cracking can occur.

For most normal external walling in the UK, bricks of 'ordinary' quality to BS 3921 should not be damaged, provided that, in winter work, they are not laid in a saturated condition. In severely exposed areas, local experience will provide the best guide. 'Special' quality bricks to BS 3921, however, may be taken to be immune to attack in such areas. Special care in selection is necessary when bricks are imported into a new area, for example, if English bricks are used in Scotland.

Local experience and advice from specialist agencies, such as the Building Research Establishment, are the best guides to the selection of frost-resistant stone. The problem is a minor one and, in general, affects limestones and not sandstones or the stones prepared from igneous rocks.

The minimum quality for mortars for masonry is specified in BS 5628 Part 3. For external walls, a mortar consisting of a cement:sand mix of volume proportions 1:5 to 6, together with an air-entraining plasticizer, should not be damaged by frost. It should be noted that calcium chloride does not prevent frost damage to mortars and should not be used.

When damage has occurred, the most likely effect is to the appearance rather than to the stability of the wall. The individual bricks or stone blocks affected may be cut out and replaced, and mortar can be repointed, using a stronger and air-entrained mix. This is all that is usually necessary but, if many bricks are involved, it may be cheaper to cover them with tile hanging or weather-boarding, or to render on lathing.

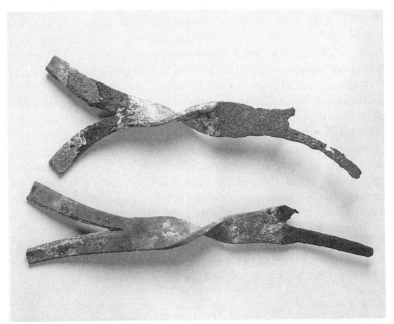

Fig. 7.9 Corroded wall ties.

Expansive damage can occur also when ferrous metals embedded in masonry, rust, and the effects on stone have been mentioned already in Chapter 3. Damage to stonework through corrosion can be prevented by ensuring that the metal fixings used are of copper, phosphor bronze, aluminium bronze or an appropriate austenitic stainless steel (see BS 5390).

In brickwork, wall ties are used to provide structural interaction between the two leaves of a cavity wall and these are usually of galvanized steel. It has been stated already that zinc has corroded rapidly when embedded in black ash mortar. This affects, principally, that part of the tie in the outer leaf. Following loss of the zinc, the mild steel corrodes, and ties of the vertical-twist type contain sufficient steel for the formation of the rust to cause the external leaf of the brickwork to expand vertically. The total expansion in a two-storey wall may be 50 mm. This results in horizontal cracking of the mortar joints in the outer leaf, coinciding with the level of the ties, usually four courses apart, and with cracks several millimetres wide. This cracking contrasts with that due to sulphate attack by being at these regular course intervals. There are usually no vertical cracks. In severe cases, the inner leaf may also show cracking and structural integrity can be impaired. Calcium chloride in cement/lime/sand mortars can also promote premature corrosion of ties. It is better not to use aggressive mortars but, where this cannot be guaranteed, it is advisable to use non-ferrous ties or those of stainless steel. However, even in normal cement mortar, corrosion can occur and cause cracking in walls which are severely exposed and if sub-standard ties are used. All metal ties used should conform to BS 1243 [47].

The action necessary following observation of the defect may vary from doing nothing to complete rebuilding of the outer leaf and replacement of all the old ties. Much will depend on whether the damage affects mainly appearance or stability. Intermediate courses of action include piecemeal replacement of corroded ties by more resistant ones and the application of cladding to reduce the wetting of the wall by rain. A proprietary foamed plastic injected into the cavity wall is also claimed to be a remedy, as well as providing some valuable thermal insulation, but long-term experience of its behaviour is at present lacking.

7.7 PROBLEMS WITH RENDERINGS

Sulphate attack and its effects on renderings are considered in Section 7.5. Renderings may crack, craze and become detached from their background, however, even in the absence of sulphate attack, through differential movements between the background and the rendering. Cement-based renderings shrink as they dry, particularly the strong, dense mixes such as a 1:0.25:3 cement:lime:sand mix. The effect of this shrinkage of renderings on rigid backgrounds is to set up stresses which may be relieved by cracking or by loss of adhesion. The cracks may then allow rainwater to pass through them but prevent its ready evaporation afterwards, which can assist sulphate attack, as already described. Cracking from shrinkage, and in the absence of sulphate attack, is likely to be random and, if tapped, the rendering may sound hollow. Differences in drying shrinkage characteristics between the top coat and the undercoat of renderings can also lead to defects. If the top coat is richer in cement and, thereby, stronger, with a greater drying shrinkage, it can pull away from the undercoat over small or large areas. When the undercoat is considerably weaker, parts of it may come away with the top coat.

Failures can be avoided by carefully following the recommendations made in BS 5262. For most situations, mixes equivalent to a 1:1:5 or 6

Fig. 7.10 Random cracking of rendering through excessive drying shrinkage.

and 1:2:8 or 9 cement:lime:sand will be satisfactory, if it is ensured that the final coat is not stronger than the undercoat. Detailed recommendations are given in Tables 1 and 2 of BS 5262, which specify suitable mixes for renderings in relation to different backgrounds, exposure conditions and the texture of finish required.

Remedial work, when cracking does not extend into the background, necessitates the cutting out of hollow areas adjacent to the cracks and patching with a suitable rendering mix. Where cracking extends into the background, the latter will first need making good, or it may be possible to fix a waterproof lathing to the background and then to render. Flaking due to the use of too strong a finishing coat is likely to need the complete removal of the finish, together with any adherent undercoat, and its replacement by a mix no stronger than the undercoat. The surface of the latter may need to be roughened and its suction reduced by the prior application of a spatterdash mix.

When corner beads are used to provide good edges and corners for a rendering, these should be of a material suitable for external exposure, preferably stainless steel. Corrosion, followed by spalling, can occur if beads only suitable for internal plastering are used.

Failed renderings are not generally very expensive to put right but, unfortunately, this is far from being the case for failures of the many other forms of cladding considered in Chapter 8.

8

Cladding

Scarcely a week goes by without accounts in the technical press, and in the columns of local newspapers, of expensive repairs needed to cladding on high-rise buildings. It is not uncommon for costs exceeding £1 million to be quoted as the sum needed to rectify faults on one high-rise building: the total national cost must be very high. Many such buildings are now surrounded by scaffolding and safety netting to prevent pieces of cladding, thought likely to break off, from falling to the ground, possibly injuring pedestrians. Even a kilo or so of brick or concrete falling from ten storeys may cause more than alarm and despondency – and some panels weigh well over a tonne.

Cladding defects are caused by differential movement between the cladding and its background; by failure to allow for the inaccuracies inherent in construction; by inadequacy of the fixing and jointing methods used; and by premature failure of sealants.

8.1 DIFFERENTIAL MOVEMENT

There are several causes of movement in claddings and of the backgrounds to which they are fixed. It is the relative movement between the two which is of first importance. The fact that substantial relative movement can occur in a tall building and needs to be accommodated has, unfortunately, been overlooked in the design of many post-war buildings. The principal causes of movement are due to temperature effects, to moisture changes and to creep. Failures have been mainly in cladding over reinforced concrete structures, most probably because steel structures have no moisture movement and creep is small.

8.1.1 Temperature effects

The thermal movement of common building materials is shown in Table 2.2. Cladding, being more exposed to the weather, is likely to be

Fig. 8.1 A fine sense of timing. Falling cladding.

subjected to greater movement, either of contraction or expansion, than the structural backing and, particularly so, when the cladding is thin and the structure massive. The movement may take place rapidly or may be more seasonal in character and, as shown in Chapter 2, will be dependent not only upon the inherent nature of the material but also upon its colour and the extent to which it is insulated. For example, significant differential thermal movement can occur and cause distortion when stones of similar coefficients of thermal expansion, but of different colour, are used. Such problems have been reported when dark marble and light granite have been used in alternate strakes, and with travertines and slates of different colours. Most cladding will be subject to a temperature range in service of 70°C in the UK, but, if highly insulated on the inside, the range can exceed 100°C. Cladding is restrained to some extent from taking up the change in size consequent upon the change in temperature. This restraint causes stresses which, if sufficiently great, may cause distortion or fracture. The extent and nature of the damage caused will be dependent upon the physical and mechanical properties of the cladding, and the extent of the temperature change. It may also be affected by the time over which the temperature change occurs and the frequency.

8.1.2 Moisture effects

Cladding is fixed, in general, to steel or concrete structures. The former has no moisture movement but concrete shrinks after placing, as its moisture content slowly reaches equilibrium with the surrounding atmosphere. An irreversible shrinkage of between 0·03 and 0·04% can be expected [48] for normal gravel aggregate concretes. Exact values for concrete depend upon the composition of the concrete mix and the particular aggregates used. Concrete also possesses a reversible moisture movement of between 0·02 and 0·06% but the concrete structure is unlikely to become wet in use, for it is protected by the cladding. It is the irreversible drying shrinkage of the structure which is of principal importance and slow shrinkage will occur over a long period. The moisture movement of common cladding materials is not usually the cause of problems. The irreversible expansion of freshly fired bricks already described can, however, be a contributory cause of failure.

8.1.3 Creep

It is probable that a major cause of cladding defects has been the failure to remember that structural concrete creeps under the dead and imposed loads placed upon it; that is, a sustained load produces a permanent deformation. This takes place over a long period of time, though the bulk will occur during the first five years. Creep is a complex phenomenon and the extent of the deformation is related not only to the applied stress in the concrete but also to the time after placing the concrete, the nature of the concrete mix and the type and placing of any reinforcement. Within the normal design stresses for concrete of 20 to 35 N/mm^2 at 28 days, the average value of creep can be taken as 30×10^{-6} mm per mm per N/mm^2 [49]. Creep of concrete at, say, 30 N/mm^2 may result, therefore, in a deformation of 0·09% which, added to the irreversible drying shrinkage, can give a total shrinkage of around 0·12 to 0·17% for the structural concrete frame. Thus, the shrinkage in a 3 metre storey height could be as much as 5 mm, and 50 mm in a high-rise block of ten such storeys. Additionally, there can be a small elastic deformation of the concrete structure under load (mainly affecting beams rather than columns).

8.1.4 Effects of differential movement

The effects of a combination of drying shrinkage and creep in the structure, and of expansion in the cladding, may be illustrated by the case of brickwork cladding to reinforced concrete. Failures in such a combination have, in practice, been all too common in recent years.

The coefficient of thermal expansion of brickwork vertically is likely to be around 7×10^{-6} per degree C and the temperature range likely to be encountered, in service, with fired clay brickwork cladding might reasonably be taken as 70°C. The maximum expansion of the cladding will depend, *inter alia*, upon the temperature at which it was first placed: a maximum vertical thermal expansion of 0·05% is possible. This will act in opposition to the shrinkage of the concrete structure to which it is attached. In total, a differential movement of some 0·22% could occur – that is, around 6 mm in a storey height, from these causes. Brickwork may also exhibit an irreversible moisture expansion, as already described, and the shrinkage of concrete can be greater if shrinkable aggregates are used.

The frame of a new timber-frame house will shrink as the moisture content of the timber reduces to the level reached when the house is in full occupation. This shrinkage will be enhanced if the frame is not kept protected and dry on site. In a masonry-clad timber-frame building, differential movement will occur. Distortion and subsequent rain penetration will then take place if gaps and joints are not properly designed and executed between the masonry and the frame. Places particularly at risk are at eaves, sills, jambs and heads of openings. Flexible wall ties will need to be used between the frame and the masonry.

8.2 INACCURACIES IN CONSTRUCTION

The likelihood of failure has also been enhanced by not allowing for the dimensional deviations which occur in manufactured or site-cast products, and for the inaccuracies in setting-out and during erection. In the past, corrective measures to obtain good fit were taken on site almost as a matter of course but, with modern multi-storey buildings, this is less possible, mainly because of the greater inflexibility to manipulation of the products used.

Supporting features, compression joints, and fixings for cladding have often not been designed or placed to allow for the likely inaccuracies over the whole building, and these can be additional to the differential movement already described. A particular example is that of insufficient bearing on the floor of masonry cladding designed to oversail the floor. Cladding is used in this way, together with fired clay 'slips' to hide the floor edges, when the design requires the elevation to show as a continuous masonry face, masking the structural framework behind. Excessive stress has also been caused to supporting nibs when the heavier cladding units of stone and precast concrete have had an inadequate bearing on them. Reinforcement is often not taken out far enough into the projecting nib, which has led to shear failures. Although the faces of concrete nibs are shown generally as being square in design drawings, this is often far from the case in reality. Mortar used to dub out the nib to the correct profile will not be reinforced and does not provide an adequate bearing to the cladding units. Inaccuracies in construction have also led to the displacement of fixing devices and failure to attach cladding adequately to the supporting structure. Many fixing devices can be adjusted only within comparatively narrow limits.

8.3 FAULTS CAUSED BY MOVEMENT AND INACCURACY

Many failures have occurred because the design and construction of the structure, its cladding and the fixings have been insufficient to allow for the likely total movement. Sometimes, mistakes during construction, such as forgetting to remove packing spacers after final fixing, have also contributed. As a consequence, a compressive force has been transmitted to the cladding which has not been designed to accept it. Characteristic failures, due to this squeezing effect, on brickwork used as cladding to reinforced concrete structures, show as horizontal cracks at roughly storey-height intervals on the line of the floor slabs, though all storeys may not be affected. This cracking is frequently accompanied

Fig. 8.2 Bowing and cracking of brick cladding to an RC frame.

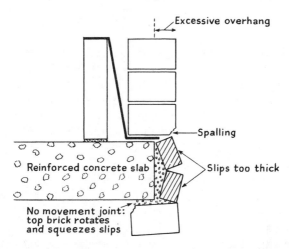

Fig. 8.3 Buckling of brick slips. Creep and shrinkage of concrete frame and expansion of brickwork causes spalling of bricks and buckling of slips if adequate compression joints are not provided.

by spalling of the edges of individual bricks in that vicinity, and buckling and displacement of a few courses of brickwork (see Fig. 8.2). Brickwork may be supported on concrete nibs projecting as part of the reinforced concrete floor slab and, although uncommon, excessive relative movement can cause cracking of the nibs, especially if they are weakened by corrosion of any reinforcement present. Where brick 'slips' have been used to cover the edge of the floor slab, these often buckle badly and spall, and may, indeed, justify their name by becoming completely dislodged (Fig. 8.3). Flexible DPCs used at floor level are often extruded. The defect is one which is unlikely to appear before several years have elapsed after construction. Similar effects can appear when natural stone or reinforced concrete is used for cladding to reinforced concrete structures and when mixed cladding materials are used. In one case reported [50], the compressive stresses generated in brickwork cladding, assisted by inaccuracies in construction, caused bending forces to be applied to an unreinforced precast stone band course, causing major cracking and failure. A further cause of failure with reinforced concrete cladding panels has been due to corrosion of the reinforcement in the panels. This has been assisted by inadequate general cover to the steel, particularly where exposed aggregate of large size has been used as the finish. In addition, compression of the edges of the panels, through the differential movement mentioned, has caused cracking and spalling and thus a reduction in the cover, leading in turn to further corrosion.

Mosaic cladding sheets may also fail from the same cause and tend to crack rather than to spall. As the adhesion between the mosaic sheets and the background is usually not strong, bulging and total detachment are the predominant defects. Ceramic tiles are commonly used for external cladding. The mode of failure due to differential movement depends to a large extent upon the type of tile. Pressed tiles lack the good undercut rear face which most extruded tiles possess and tend to become detached at the interface between the tile and the bedding mortar. The more commonly used extruded tiles, together with adhering mortar, tend to come away from the structural background. Failure of brick slips, mosaic sheets and ceramic tiles is also likely if fixing and adhesion are poor (see Section 8.5.2).

Glass-reinforced cement (grc) cladding panels are made from alkali-resistant glass fibres, usually in the range of 34–38 mm long, cement and sand. A field survey of grc cladding panels in use in the UK showed an unacceptably high proportion of buildings inspected to have some cracked panels [51]. In part this was due to some deficiencies in manufacture but also to undue restraint imposed by fixings and a failure to recognize the effects of adverse combinations of temperature and moisture change.

Glass used as lightweight cladding in curtain walls can crack through differences in the rate of response to changes in temperature between it and the framework of metal into which it fits. Metals have a higher coefficient of thermal expansion than glass, often twice as much, and may exert sufficient force on the glass to cause it to crack, if clearances between the glass and frame are too small. The crack is usually a single crack, starting from the edge of the frame.

8.4 SEALANTS

Flexible joints used in cladding are affected both by movement and by inaccuracy. The different sealants used for jointing have properties which allow certain movement to be accommodated safely in relation to the width of sealant used. This movement, expressed as a percentage of the minimum width of sealant necessary, is shown in manufacturers' literature as a 'movement accommodation factor'. The wide variety of sealants available covers a wide range of movement accommodation factors but, taking a fairly typical factor of 20%, the minimum width of sealant necessary in the joint would then need to be five times the likely total movement. There have been many failures of sealants through failure either to determine correctly the likely total differential movement or, having determined it correctly, to use a width of joint which can accommodate that movement but is, nevertheless, too narrow for the sealant chosen. A sealant unduly stressed will tend to break down and its durability will suffer. This, in turn, may enable rain to penetrate, which would not otherwise happen.

8.5 FIXING METHODS

Cladding is fixed to the structure, usually by metallic fasteners of various kinds or by adhesive.

8.5.1 Metallic fasteners

The main defects associated with metallic fasteners have been their inability to perform the design function satisfactorily and their failure through corrosion.

The inaccuracies inherent in the construction of high-rise buildings require that fixings have a fairly wide range of adjustability but many do not, and the measures taken on site to get a fit are often unsatisfactory. Common faults are the lack of full and proper engagement of dowels in the slots or holes meant to receive them; angled surfaces instead of square ones, which makes good bolting difficult to achieve; and too great a thickness of packing pieces used on bolts, which can reduce the efficiency of the latter. Sometimes, holes or slots incorporated in the fixing device, compressible washers and long-threaded bolts, all used to allow adjustment to be made, cannot be properly utilized because of constraints in adjacent parts of the construction. Typical problems have been illustrated by Bonshor [52]. More in the realm of carelessness than inaccuracy is the failure to tighten bolts properly or sometimes to overtighten them which, in the latter case, can cause excessive stress concentration on the cladding material.

Corrosion of metals is considered in Chapter 3. Cladding panels are used in exposed conditions and those based on heavy, porous materials, such as stone and concrete, tend to be on the thin side to keep down the overall weight. It may be expected that, under those conditions, the fixings used with them will be wet for long periods and in an environment conducive to corrosion.

Corrosion may be caused by well-known effects, including electrolytic attack which may occur if packing pieces of one metal are left behind to react with different metals used for the main fixings. The combined effects of corrosion and stress can cause stress corrosion cracking. Atmospheric pollution and moisture, coupled with stress corrosion, have been known to attack manganese bronze cramps used with stonework. Corrosion of fixings causes staining on the face of the cladding and the cracking or spalling of brick, stone or concrete cladding close to the points of fixing. It thus gives useful prior warning that a thorough investigation of the state of fixings is necessary to avoid complete detachment of panels. However, excessive force on non-ferrous bolts, due to lack of adequate allowance for movement of cladding slabs, has led to shearing of the bolts and sudden detachment of the slabs without such warning.

8.5.2 Adhesives

Brick slips, mosaic sheets and ceramic tiles are the commonest forms of cladding attached by mortars or organic adhesives rather than by mechanical fasteners. The differential movement of the structure and the cladding due to thermal, moisture and creep effects are described above. The shear stresses imposed on brick slips, mosaic sheets and ceramic tiles have often destroyed the bond between them and the mortar or other adhesive used, or between the latter and the structural background. It is probable that water reaching bedding mortars and freezing, in the severe exposure conditions to which claddings are often subjected, has also contributed to failures of bond with brick slips which are porous and allow water to penetrate. Inaccuracies in the nib of the floor slab to which the slip, mosaic or tile is to be stuck can lead to the need for either excessively thin or thick adhesive beds, neither of which contributes to success. Thicknesses required for bedding mortar for brick slips to keep them in the same vertical plane as the rest of the cladding have been found to range from 3 mm to more than 25 mm. The preparation of the background is of critical importance. Failures on concrete backgrounds have been assisted by the preparation and cleaning being insufficiently thorough in removing mould, oil and laitance. Dense structural backgrounds, too, have low suction, which has led to difficulties in the adhesion of mortar beds. Unfamiliarity with the many different organic adhesives available has also contributed to failure. Contributory causes have included the incorrect choice of adhesive; storage which has been too long before use or at temperatures which cause deterioration; and lack of control in the proportioning of the various ingredients which go to form the adhesive. Many organic adhesives have two ingredients which have to be mixed accurately, and in the right order, before use. Failure to do so has led to lack of early strength and a short life in service. Proper application of the adhesive is required: often too thin a coat has been used or the adhesive has been dabbed on at a few points rather than spread completely and evenly over the cladding unit.

For brick slips to floor nibs or edge beams, the three adhesives mainly used have been based on epoxy resins, polyester resins or sand/cement mortars modified with styrene/butadiene rubber. Epoxy resin adhesives are two-part, or sometimes three-part, adhesives which need to be carefully proportioned, mixed in the right order and, to be effective, applied to a thoroughly clean but roughened surface: support is also needed for up to twenty-four hours. These are not properties to which normal site working and workmanship are sympathetic. Polyester resin systems vary considerably in their properties, particularly in their working life after mixing and in their ability to bond successfully to damp surfaces. The ratio of resin to hardener is often critical. The sand/cement/styrene/butadiene rubber adhesives are closer to normal mortar mixes in their preparation and application and can be used on damp surfaces. Their strength development is low at low temperatures. Failures using these adhesives have been caused by too thick an adhesive bed due to inaccuracies in the positioning of the concrete floor nib; too smooth a face to the floor nib; lack of adequate grouting of the brick slip and of the nib; and inadequate coverage of the slip by the adhesive. Delay in positioning the slips after applying the adhesive can also lead to detachment. Failure from this delay may be assumed when the pattern of the adhesive formed by using the customary notched trowel can be seen to be still undisturbed on the failed slip.

Mosaic sheets come in two main forms: paper-faced mosaic in which the pieces of mosaic are glued face down to paper, or bedding-side down to nylon strips or nylon fabric (nylon-backed mosaics). In the latter case, the backing is embedded in the mortar or adhesive. Failures occur from the same general causes as with brick slips. Additionally, if final straightening of the joints is done after the bedding has started to set, the latter can be stressed and the bond broken between it and the mosaic. It is a mistake to suppose that, because the face of the mosaic is relatively impervious, rain will not reach the back. The large number of joints between tesserae allows rain to penetrate and pass into the building mortar or adhesive, from which position it can be slow to escape. This will enhance the possibility of frost and sulphate attack if mosaics are applied to brickwork containing soluble sulphates. There have been spectacular and costly failures of mosaics from the latter cause.

8.6 PREVENTION OF LOSS OF INTEGRITY IN CLADDING

Ways of reducing the likelihood of failure are fairly self-evident from the foregoing sections. The first essential is to be aware of, and to calculate, the extreme range of differential movement likely to occur between cladding and background, and to ensure that the compression joints and movement joints required can cater for this, as well as for the inaccuracies likely in manufacture and construction. Guidance on the latter is available in BS 5606 [53]. The width of joints will, in addition, need to be sufficient to allow movement so as not to cause undue stress to the sealant to be used, and reference will need to be made to manufacturers' information, in particular, to the movement accommodation factors. Compression joints will also need to be wide enough and filled with materials which do not compress so much that unwanted forces are transmitted to the cladding. Minimum compression-joint thickness needed per storey height is likely to be around 12 to 15 mm for brick or precast concrete cladding applied to a reinforced concrete background, but these values are guides only and should be considered as probable minima in normal construction. When the cladding is to be supported directly by the structural frame, it is essential to provide a bearing of at least 50 mm after allowing for all the dimensional deviations probable. The structural consequences of the failure or omission of one or more fixings should be evaluated. Where precast reinforced concrete cladding is used, cover to reinforcement should be a minimum of 25 mm and, generally, in accordance with Table 1 of BS CP 116 [54]. Where exposed aggregate finish has been used containing stones of 75 mm or over in size, the reinforcement should never be less than 20 mm behind the stones. Methods adopted for fixing claddings to backgrounds are many and, often, complex. It is essential that, at design stage, careful thought is given to the inaccuracies likely to arise in construction and the effect these will have on the ability of the fastening device to hold the cladding securely. These inaccuracies are likely to be greater than the designer anticipates and he should consciously increase his own estimate of them. Fastening devices should be chosen which have a high adjustability and, in design, provision for adjustability should be considered separately from provision for movement during service. The aim should be for an adjustability which can be made easily on site and for the integrity of fixings to be capable of being checked before being covered by the cladding. The fewer the fixings necessary the better. Portions of fixings in the structural frame need to be positioned with great care, which implies a high degree of site control. The critical alignments and dimensions of fixings, and of sockets in particular, need to be specially identified and quantified in design, and to be made known to those involved in the construction process. No changes should be improvised on site without prior reference to the designer. Useful guidance on fixing problems is given in BRE Digest 223 [55] and 235 [56] and by Bonshor and Eldridge [57] in their work on tolerances and fits.

Designers should assume that fasteners in contact with cladding will get wet and, consequently, should be chosen to remain free from corrosion under these circumstances. The possibility of bimetallic corrosion, particularly if there is any mild steel involved in the construction, should be evaluated. Metals shown in BS CP 297 [49] should be suitable for fasteners. They include phosphor bronze, silicon aluminium bronze, copper and certain types of stainless steel. The use of ferrous metals is permitted in that Code but with a qualification which seems to be of doubtful real value. It would seem better to avoid the use of ferrous metals altogether in view of the danger if fixing devices fail, and the high cost of remedial works. It would also be desirable for the design to incorporate ways by which the condition of any fixings can be inspected during their lifetime without undue disruption and difficulty.

The best advice on brick slips is not to use them but, if aesthetics overrules prudence, then the designer and client should understand that the adhesives used so far cannot be guaranteed to be successful. Much will depend on the specific design, but even more on the site workmanship and control, the choice of adhesive and its proper mixing and use. Failure sooner or later is more likely than success. Success may be more probable, however, if some form of mechanical anchor is used instead of, or in addition to, adhesives, though these may spoil the unbroken line of the slips.

To have much chance of success with mosaic and ceramic tiling the detailed recommendations given in BS 5385 [58] need to be followed carefully, as do the precise recommendations of the manufacturer of the adhesive when organic adhesives are to be used. These will usually cover such items as the type of trowel to be used, the mixing procedure, the working time after the adhesive is spread and the suitability of the background. As with brick slips, failure is, nevertheless, more probable than success and there are more constraints to using mechanical support systems.

To minimize the risk of cracking of grc the coefficient of thermal expansion should be taken as 20×10^{-6} per degree C, the reversible moisture movement around 0.15% and the irreversible drying shrinkage 0.05%. There will be less risk of cracking if small, flat sheets are used than larger panels of more complex section.

Where glass is used for cladding, the most common cause of failure can be prevented by ensuring that clear glass has a clearance all round of at least 3 mm. Coloured glasses, or clear glass with a dark background close by, can reach temperatures as high as 90°C even in the UK. Greater movement must be allowed for – not less than 5 mm where the longer dimension of the glass exceeds 750 mm.

It is not useful to generalize on the remedial measures necessary once cladding has failed. Many variations of systems of cladding and structural frames are possible, and failures have ranged from those where complete replacement has been necessary to those where local patching has sufficed. It is true to say, however, that most cladding failures are highly expensive to rectify. The need to estimate properly the likely differential movement and inaccuracies in construction, and to allow for them adequately, needs special emphasis. So does the need to ensure that reinforced concrete cladding panels have wholly adequate cover to reinforcement, particularly if an exposed aggregate finish is used.

8.7 WATER ENTRY

Many cladding systems have proved incapable of preventing the entry of rain. In part, rain penetration has followed the cracking, spalling and dislodgement of cladding caused mostly by the differential movement described. Lack of attention to joints and to rain-shedding features has, however, also been a cause.

There are many forms of joint associated with the various claddings used, some complex in geometry (which adds to the likelihood of difficulties in site work). Common problems of rain penetration, however, may be illustrated by reference to three main classes of joint – the butt joint, the lap joint and the drained open joint. The butt joint, formed when units are simply butted together, is the simplest and relies on a gap-filling material, usually a mortar, organic sealant or gasket, to permit movement and to keep out the rain. The gap-filling material has to take the full movement of the joint. Cement-based mortars shrink and their adhesion to most backgrounds is not particularly good: generally, one would not look for their use with large cladding units. Failures with organic sealants have been due mainly to their use in butt joints which are too narrow and too deep, leading to undue stressing, as described in Section 8.4; this, in turn, has led to breakdown. Inaccuracies in panel manufacture, particularly in joint width, and in fixing on site which has led to panels being out of true, have contributed to failure. Inadequate cleanliness and inappropriate surface texture of the cladding have led to loss of adhesion and allowed leakage.

When cladding panels lap across one another, the lap joint formed imposes less strain on the sealant and this is also partially protected from the weather. Failures have been due largely to lack of good adhesion of the sealant to the surfaces of the cladding units. Gaskets are pre-formed materials based on rubber or plastics and depend not upon adhesion but on being squeezed by the units to provide the seal. They are used mainly with large cladding units, particularly pre-cast concrete panels, in drained open joints but may also be used in lap joints. Gaskets cannot adjust well for irregularities in joint width, or for excessive roughness of the surfaces to be joined. The joint at the intersection of vertical and horizontal gaskets can be difficult to design and achieve on site.

The drained vertical joint consists generally of an outer zone, essentially to trap rainwater and provide drainage, and an inner zone which provides an air-tight barrier. This also is intended to prevent entry of any rain which has penetrated past the outer zone and the baffle which is commonly provided (Fig. 8.4). The materials used for baffles need to have a long life, for they are usually difficult to replace. The air-tight barrier is formed by a sealant or gasket and is not exposed to weathering. This inner barrier is essential to the success of the joint and

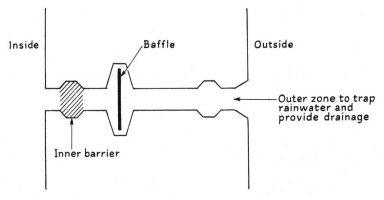

Fig. 8.4 Plan of a typical drained vertical joint. The baffle is often difficult to replace.

needs to be air- and weather-tight to prevent any water which reaches it from passing by capillarity into the building. Drained open joints have flashings provided to direct to the outside, any rain which has passed the baffle. Horizontal joints are usually protected by an upstand, so that rain cannot blow directly through the joint. This, generally, takes the form of the lower edge of the upper unit projecting downwards to some 50 mm below the top of the upstand of the unit below. Wind and rain conditions around high-rise buildings, in which large-panel construction using drained open joints is mostly used, are severe, and penetration of rain has occurred through failure to make the internal air seal effectively, because of difficulty in access to the back of the joint; by failure of the flashing to bring the rainwater to the front of the joint; by omission of the flashing and inadequate lapping and sealing; by inadequate depth to upstands; and by baffles at the intersection of vertical and horizontal joints failing to reach down far enough, thereby reducing the effectiveness of the upstand. Many of the difficulties have occurred through failure to recognize the tolerances inherent in manufacture and the inaccuracies in construction.

Weatherstrips or throatings are sometimes used as horizontal projections, their purpose being to shed rainwater clear of the units below. Failures have happened when they have collected rainwater which has been blown by high winds through the joints.

When water does appear at the inside of a building, the path of its entry may not be obvious, and a detailed and costly examination is often necessary to identify the weak points in the cladding system. Water appearing at a low level may well have drained from above or, in some forms of curtain walling, have travelled considerable distances through hollow sections. In curtain walling, wind pressure on glazing can cause displacement of glazing compounds when distance pieces are omitted or too narrow. This has allowed rainwater to be pumped past the compound, especially at high levels. Distance pieces are commonly made of plasticized PVC and are resilient. They need to fit tightly between glass and frame. Design recommendations are given in BS 6262 [59].

Water penetrating through cladding can be difficult to trace and expensive to rectify. With heavy cladding panels, it may, indeed, never be possible to gain access to, and replace, sealants, gaskets, baffles and flashings. Problems arise, in the main, at junctions of cladding units with one another and at openings. Design detailing here needs much care and should not be left to inexperienced designers. It must take into account tolerances and site inaccuracies and be as simple, geometrically, as possible. Design should strive to allow reasonable access to key parts of the water-barrier system for even the most durable sealants, corrrectly used in the proper joints, are likely to have a life expectancy less than that of the building of which they are a part. Close liaison between designer and contractor on the feasibility of joint construction in relation to the proposed design is vital. Jointing is not something which can be left to the site operative to fudge as best he can.

9

Doors and windows

Decay of timber external joinery by wet-rot fungi has been widely reported in recent years and has affected many comparatively new properties. Both rain and internal condensation have been the sources of moisture which caused decay. Rain penetration at window and wall intersections has been not uncommon.

9.1 DOORS

A wide survey of decay in doors, carried out in the UK, showed that a high percentage of unprotected external panelled doors, aged twelve years or less, showed decay [60].

Entry of moisture has occurred mainly at joints and, with glazed doors, has soaked into the framing where putty or glazing beads have come away from the glass. Such beads frequently offer little protection. A contributory cause of rain penetration has been inadequacy of the tenons of the bottom rail, leading to slight dropping of the rail and consequent exposure of the joint between the rail and the panel. The absence of a weatherboard on the outer face at the bottom of the door has meant that rain is not thrown clear of the gap under the door. This problem has often been made more serious by the omission of the weather bar from the sill or its incorrect positioning. Entry of water at the joints can lead to failure of the urea formaldehyde glue usually employed. It can also promote differential moisture movement where the grain of the timbers which meet at the joint runs in different directions. The stressing of the joint through moisture movement, and its weakening through failure of the glue, leads to loosening of the joint and an increased opportunity for moisture to enter and cause decay. Delamination of plywood panels in external doors has occurred when internal-grade plywood has been used. The outer ply becomes wrinkled and often shows signs of splitting at the edges. Plywood, no matter what its grade, is at risk from decay if moisture can enter through unpainted

Fig. 9.1 Wet rot in window frames.

end-grain. Distortion of doors is also likely if moisture penetrates and may occur, too, if humidity and temperature conditions are markedly different on the two sides, as they often are for an external door, and the door is poorly protected from such changes by the lack of an adequate paint treatment. Doors may then jam or give a poor fit in the door frame.

9.2 WINDOWS

Similar, but more extensive, problems have occurred with many timber windows. Modern machine-made joints are more complex than the older mortice-and-tenon joints, which were primed before assembly, and often result in large surface areas in contact. The adhesive used is not always applied in sufficient quantity, and with sufficient control, to ensure that all such surfaces are properly covered, and some are left as comparatively easy routes for moisture to penetrate. In any case, the adhesive does not seal the surfaces as effectively as primer. Some window frames used have been insufficiently robust and have distorted

Fig. 9.2 Rain penetration at window/wall joint.

through moisture movement, this distortion then permitting the easier entry of moisture.

Moisture may find its way into window frames for a variety of reasons. Unfortunately, apart from the modest amount given by the manufacturer's primer, protection during transit, on site, and after installation, is frequently poor or even totally absent. The moisture picked up by timber windows as a result can be trapped when the frame is painted prematurely. Moisture may also enter through contact between frames and wet masonry. Current practice is to set the window frame in the outer leaf, which will be wet for long periods of time, and the frame itself, being so far forward, will also be more exposed to rain and sun. Rain may find its way directly behind glazing putties, which have moved away slightly from the glass, and into the joints of joinery when these have not been properly protected by paint. Rain will also collect on the horizontal surfaces formed if top-hung or centre-pivot lights are left open. An important route for entry, and one which seems to be increasing, is through condensation on the inner side of the glass. Condensation has been dealt with in some detail in Chapter 4. It may suffice here to state that condensation running down the pane collects on the usually rough and horizontal surface of the back putty and may remain there for long periods, slowly finding its way into the frame. Once it has entered, subsequent evaporation may be slow especially when, as is usually the case, both the internal and external faces of the frame are painted with a relatively impervious high-gloss paint.

9.2.1 DPC at window openings

Rain penetration at the junction of window frames with walls has been due to inadequacies in the detailing of the DPC at jambs, heads or sills. Windows are usually positioned as the wall is being built and this is preferable to fixing the window in an already formed opening, for it is then much more difficult to ensure an effective barrier to rain. The window frame can be used to close the cavity but usually the frame is set towards the external wall face and the internal leaf of the wall returned to close the cavity. Failure commonly takes place at the jamb through reliance on a simple butt joint between the vertical DPC and the frame, and also through failure to extend the DPC into the cavity (Fig. 9.3). A simple butt joint is most unlikely to provide an effective barrier, even in sheltered conditions. Any shrinkage away from the wall will leave a gap for rain to pass directly between the frame and the wall, if the junction there has not been well pointed with a mastic sealant. Even if it has, the

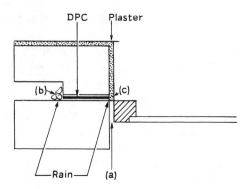

Fig. 9.3 Rain penetration at window jamb. Poor practice: (a) gap, caused by shrinkage of frame from wall, not sealed with mastic. (b) DPC, not projecting into cavity, can be bridged by mortar droppings. (c) DPC·bridged by plaster if inside face of frame is flush with inside face of external leaf.

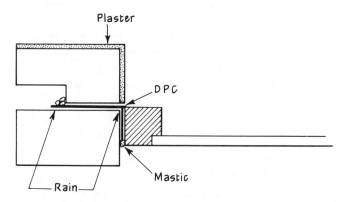

Fig. 9.4 Better practice. DPC projects at least 25 mm into cavity and is tacked to the window frame. Mastic adds to the weatherproofing.

DPC can be bypassed if the construction is such that internal plaster can be directly in contact with the external brickwork at the jamb. This can happen when the inside face of the frame does not extend beyond the inside face of the external brickwork. Rain penetration from this cause will be particularly likely if strong, relatively non-porous bricks, unable to absorb rainwater, are used for the external leaf. Wrong positioning of the vertical DPC with respect to the frame, allied to lack of, or inadequate, mastic pointing, is a common cause of rain penetration and one which should be suspected as the most likely when it occurs. The details to be recommended will vary with the type of window, its relationship to the face of the wall and whether it is to be built in as construction proceeds or placed later in an opening. A positive seal can be achieved by having a recessed, or grooved, frame and projecting the DPC into this. The frame can also be kept out of direct contact with the brickwork. The DPC should then be nailed or stapled to the frame before the window is built in. The DPC will then need to be wider than the normal half-brick width, but a wider DPC is needed, anyway, to ensure that it extends at least 25 mm into the cavity to prevent its being bridged by mortar droppings. Where joints in vertical DPCs are necessary, the upper piece should lap to the external face of the lower piece. Care should be taken to see that any mastic used for sealing can take the strain which will be caused by differential movement between the window frame and the wall. The correct joint width for the sealant used needs to be achieved (see Chapter 8). If the joint to be sealed is open at the back a back-up strip, usually of foamed plastics, needs to be used. This will help to ensure that the mastic sealant is forced against both sides of the joint and thus gives a good seal.

It is common practice to stop sills just short of the wall and, in theory, this could be useful, as it prevents the ends from being in direct contact with wet masonry. However, if the joint is unfilled, rain can easily penetrate it and may possibly reach the interior. It will undoubtedly reach the end-grain of timber sills. The joint should be well filled with mastic. Rain can also penetrate at the bed joint beneath the sill when the latter does not have an adequate projection beyond the face of the wall, when the slope is shallow and when the drip is poorly formed or omitted. Many timber sills are constructed from two separate pieces, for this enables smaller and, thus, less costly, pieces of timber to be used. It is essential that the outer piece sheds rain quickly and in no way enables water to be retained or fed back to the inner section. Examples are frequent where this has not been the case.

Masonry or rendering underneath a sill will be damp and may be excessively so when the sill projection is inadequate. Good practice requires a DPC below the sill to prevent such dampness from rising to reach the sill and this will need to be detailed carefully to link with the vertical DPC at the jamb. The sill DPC will need to be turned up to form an upstand and the jamb DPC to overlap it so that rain is shed out and not in.

The lack of condensation channels on the inside surface of window sills has been lamented already. The need to slope the surface of the sill here is as necessary as it is on the outside, so that condensation does not feed into the window assembly, be it timber or metal. Too often, nowadays, the sill detail approaches that shown in Fig. 9.5 rather than that desirable (Fig. 9.6).

Setting windows back from the face of the wall helps in giving the window assembly some protection from the weather but will usually necessitate the use of a sub-sill to throw rain clear of the wall. The top surface of the sub-sill will receive run-off from the glazing and will itself be exposed to the weather. Water penetration between the sill and the sub-sill can occur unless detailing is good and construction properly undertaken. The use of a weather groove in the sill is a common method of preventing such penetration but this needs to be positioned accurately so that any water penetrating drips into the cavity of a cavity wall and not straight onto the inner leaf.

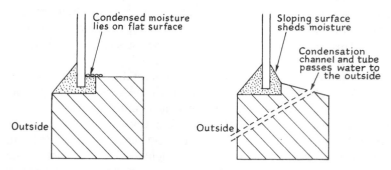

Fig. 9.5 Damage by condensation on inside of window frame.   **Fig. 9.6** Removal of condensed water.

The causes of rain penetration at window heads are similar to those dealt with in Chapter 7. Failures have occurred through inadequate lapping of the materials used in the construction of the cavity tray at the window head and by damage to it. Mortar droppings, too, have led to bridging of the DPC and to restriction of discharge of water through weepholes. Poor detailing at the junction between head and jamb has allowed rain to penetrate to the inner leaf at the ends of the lintel. Where the inner leaf is of block, failures have occurred because the normal depth of flashing of 150 mm is insufficient to reach, and to be tucked into, the first blockwork joint. This difficulty can be overcome by using either a flashing of greater depth or shallower blocks just above the lintel. The cavity tray should extend beyond the edge of the jamb DPC and the toe of the tray should project past the window head.

The pressed steel lintel, in most cases, acts also as the cavity tray and is available in lengths which obviate the need for jointing – always a

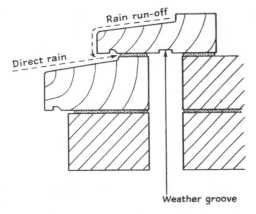

Fig. 9.7 Sill must be accurately placed so that weather groove is over the cavity and not the inner leaf.

source of weakness. The lintels are of galvanized steel and other protective coatings may also be applied: so far, chlorinated rubber paints and thermosetting epoxy powder coatings have been used. Provided that the long-term durability of the protected steel is satisfactory, this development seems to have promise in helping to prevent many of the problems associated with the separate introduction of a flexible cavity tray at window heads. The use of steel lintels may give rise to cold bridging and condensation (see Chapter 7), but some are designed to be plastered with an insulating plaster on the portion which is supported by the internal leaf and this should go some way to mimimizing the difficulty. Metal window frames have a high thermal conductivity and provide a cold bridge between exterior and interior. This can lead to condensation problems, too, which can be quite severe. Insulation needs to be added to the frame to reduce the problem and this can be nearly impossible after installation. It may be noted that, while double glazing can reduce the risk of condensation on the glass itself, it will not necessarily prevent all condensation there: this will depend upon the internal humidity and temperature. Double glazing will not, of itself, affect the likelihood of condensation on the frame.

9.3 PREVENTION OF FAILURE AND REMEDIAL WORK

The presence of decay in the woodwork of doors and windows is readily detected. The timber surface is often dished, through the falling away of the soft and friable underlying timber. Paint surfaces are generally cracked and peeling away from the underlying wet timber. Putty is usually loose, having shrunk away from the timber, and pieces are cracked or missing. Replacement of putty is comparatively straightforward but, whether or not it is feasible to cut out any decayed timber and renew it with preserved timber, clearly will depend upon the extent of the decay. Total replacement may be necessary. Where decay is localized, it may be possible to cut out the affected wood, to apply liberally a suitable preservative and to replace with treated wood. It will be necessary, generally, to remove and replace the putty and paint, and to fill open joints with a water-insoluble filler. The British Woodworking Federation has its own performance standard for wood windows and included in the scope of this standard is the requirement for preservative treatment where necessary. In practice this means that all softwood windows manufactured by its members are impregnated with preservatives.

Where DPCs and cavity trays are defective and have allowed rain to penetrate, the inner walls will show local patches of damp after rain, adjacent to the defect, though these may be masked by decoration and the first signs of trouble may be after drying out, when efflorescent salts may appear. Remedial work is likely to be difficult and expensive, particularly to DPCs and cavity trays at jamb and head. Cleaning out mortar droppings from a flexible cavity tray after the mortar has hardened may be almost impossible without damage to the tray. Rain penetration may often be prevented or minimized by mastic pointing. Pointing is a simple operation and should be tried before replacement of existing DPCs is contemplated. It may also be possible, in areas where exposure is slight to moderate, to use a water repellent, despite the possible drawbacks mentioned in Chapter 7.

Specific detailed checks are strongly recommended at all parts of the DPC system before the work is finished. The cost of checking the initial design of DPCs at window openings and, carefully, the adequacy of site construction, is a small fraction of that which will be necessary to open up and replace the DPC later. This is true even at ground level in simple dwellings, let alone for more complex buildings at a height.

A thermal expansion problem can arise with windows made from uPVC. Dark colours can give rise to dimensional instability and enhance the risk of cracking of the glass. This is because the material has naturally a high coefficient of thermal expansion and dark colours will lead to greater temperature changes. It is safer to choose white frames and to keep them clean to avoid the build-up of dirt.

Finally, it should be noted that external doors and windows need to be selected to fit the probable exposure conditions and not bought on grounds of lowest possible price. Windows, in particular, are very exposed and often not easy to replace. A little extra cost here could be money well spent, though even a better-quality window will need far better handling and storage on site than has been customary in recent years.

10
Roofs

As a generalization, it can be said that pitched roofs have given few problems and flat roofs have given many. The best advice one can give is to use a pitched roof wherever possible. However, post-war design has resulted in many buildings, particularly offices and schools, being of wide span, which has undeniable advantages. To top a wide-span building with a pitched roof would often mean that the total roof height would be unacceptable both in terms of visual appeal and first cost. For such a building, a flat roof may be the only feasible solution. To use flat roofs for buildings of comparatively small span, however, such as dwellings, is to court disaster, as many local authorities have found to their cost. The chances of success over a long period are minimal, given current standards of construction and the heating and ventilating regimes likely in dwellings.

10.1 FLAT ROOFS

Flat roofs fail because they let rain through, construction water is trapped which afterwards drips out, or moisture generated within the building condenses and drips back. Moisture from the last two causes can also assist in the breakdown of the waterproof covering which, in turn, can lead to rain penetration. With some roof decks, the continued presence of moisture has been a contributory cause of structural failure. A major cause of leakage has been an insufficient slope to the roof. A designed fall of 1 in 80 does not result in a finished fall of that gradient. Variations in constructional accuracy, settlement, and thermal and moisture movement, lead to lower finished falls. Local areas of the roof, too, can deviate markedly from the overall falls, particularly around features such as roof drains. Local ponding, rather than shedding, of rainwater then occurs, which in turn can lead to further local deflections, more ponding and slower drying (see Fig. 10.1). Rain leakage through roofs following splitting and deterioration of the waterproof

Fig. 10.1 Ponding on flat roof – a state typical of many flat roofs.

coverings has been due, in part, to failure to recognize, or to allow for, the differential moisture and thermal movement to which a flat roof system is prone; to the adverse effects of standing water and solar radiation; to blistering caused by pressure of entrapped moisture or air; to inadequate detailing at parapets and projections; to decay and collapse of some materials used as decking or for insulation; and to mechanical damage. If a decision has been taken to use a flat roof, then the designed falls should be 1 in 40 and preferably achieved by sloping the structure rather than by forming them by a screed or in the insulation.

There are many types of flat roof and waterproof covering. There are, however, some issues which are common to much flat roof design and construction, and these are now considered.

10.1.1 Dripping of moisture

Water used in the construction of concrete decks, and in lightweight concrete screeds used to provide falls and as thermal insulation, can be slow to evaporate. In the UK, an unprotected deck or screed is unlikely to dry out, except in a long spell of dry, warm weather. Only in the driest parts of the UK, in fine weather, does evaporation exceed average rainfall and it is seldom possible to apply waterproof coverings to dry concrete roof systems. A common failure has been the dripping of water, usually stained brown, from the underside of the structural roof and, particularly, from natural outlets, such as electrical conduits. Very often, the cause of this dripping is taken to be rain passing through a defective covering. This may be the case but it is more likely to be caused by entrapped water used during the construction of the deck or by entrapped rain which fell during construction. As mentioned, most roof-laying operations in the UK will be affected by rain. In some cases, this may not matter, because the water will drain through the roof and evaporate more or less harmlessly from the underside. However, when a vapour barrier is used beneath the roof deck, rain may be trapped between the vapour barrier and the waterproof covering. It can remain trapped for some years, very slowly migrating towards low points in the structure and emerging wherever it reaches some feature which has resulted in a discontinuity in the vapour barrier. It then drips out. When a vapour barrier has not been used, moisture derived from internal activities may condense beneath a waterproof covering, particularly, though not exclusively, in the winter and can drip back in a similar way.

It may be difficult to decide the exact cause of dripping, but that originating from trapped rain or construction water will have no marked seasonal tendency to manifest itself, while that due to condensation is more associated with the winter. If it has been through leakage of the waterproof covering, then it will be associated with rainy periods though not necessarily coincident with the rain. The position at which drips appear can, of course, give some indication of the cause. Thus, defects in parapets, skirtings and verges are likely to result in drips near the junction of the ceiling and external walls. Defects caused by movement at construction joints may well give rise to drips near the junction of ceilings with internal loadbearing walls.

Any entrapped moisture needs to be drained away. Drainage can be effected by puncturing the underside of the roof at the low points. The holes are closed later when the base has dried. Ventilators passing through the waterproof covering may also be of value in helping to release moisture-vapour pressure, though ventilators will not, of themselves, dry out the screed or base. No vapour barrier should be used which could seal in such water.

10.1.2 Warm-deck and cold-deck roofs

Condensation in flat roofs, whether they are covered with asphalt or are constructed in other ways, can be prevented or reduced by providing an effective vapour barrier at the warm side of the roof structure and by ventilating the roof to the outside air. In the case of a flat concrete roof, condensation can be prevented, in general, by the provision of insulation above the structural deck and separated from it by a vapour barrier. The latter should be of high quality: the view has been expressed [61] that bitumen felts types 1B or 2B of BS 747 [62] are scarcely adequate and some holding Agrément certificates have been preferred. The roof described is known as a warm-deck roof because, in a heated building, the roof deck will be warmer than it would be if the insulation were provided underneath the deck. With insulation above the deck, the latter, and the ceiling below, will be at a temperature close to that of the interior of the building. The insulation is generally protected at its upper surface by the waterproof covering. In a warm-deck roof, it is not necessary to provide cavities in the roof system specially for ventilation except in cases such as heated swimming-pools where pressurization with fresh air may be necessary. However, a warm-deck roof will have a cavity formed between it and the ceiling when this is of plasterboard on battens. Such a cavity is of value if the building is intermittently, rather than continuously, heated. If the insulation to be used is, itself, not affected adversely by water, either in relation to durability or to the retention of its insulating properties, then it may be placed not only above the deck but, also, above the waterproof layer. The commonest of such insulating materials are foamed glass and extruded polystyrene. This type of roof construction is, commonly and rather confusingly, termed 'inverted' warm-deck roof construction. A better name, also used, is a protected membrane construction. Both the protected membrane construction and the conventional warm-deck roof construction will pose different constructional problems. In the case of the protected membrane construction, the waterproof layer is protected from sunlight, from large thermal movement and from any traffic on the roof. Some limited experiments over a six-month period have shown that temperature fluctuations in asphalt can be as little as one quarter that of asphalt placed over the insulation and the rate of temperature change very much slower [63]. A protected membrane roof also avoids the need for a separate vapour barrier, as the waterproof layer performs this function itself. Construction moisture is not trapped and can dry out downwards. However, the insulation will need to be held down against wind forces, drainage design will be more demanding and the waterproof layer may not be accessible to any repair needed without partial destruction of the insulation. It will be necessary to weight down the insulation by gravel, by slabs or some other means as the work proceeds to prevent it being blown away. In conventional warm-deck roofs, where the insulation is below the waterproof covering, the latter is readily accessible for maintenance but is subjected to much thermal movement. If leakage does occur, water may be trapped in the insulation and this can reduce its insulating properties. It can cause rotting if the insulation is cellulosic. A separate vapour barrier is required below the insulation. Good solar protection to the waterproof covering will also be needed. The covering will also need to be protected against thermal movement at the joints between rigid plastics insulating boards.

In cold-deck roofs, thermal insulation is provided below the roof deck. A ventilated air space will be needed between the ceiling and the deck to reduce, or prevent, condensation, and this can pose many difficulties in design and construction. While it may be possible to design and construct cold-deck roofs to prevent general condensation it is difficult to prevent localized condensation and the risk of failure is greater than for warm-deck roofs. Protected membrane construction offers the balance of advantage for heavy decks of concrete. It may not, however, if it is of lightweight construction and particularly when it is of troughed metal. It has been shown [64] that heavy precipitation during cold weather can reduce the effectiveness of the insulation and can lead to a significant fall in temperature at the underside of the deck. This may be sufficient to cause condensation there. This risk should be considered consciously before the decision is taken to place the insulation above a lightweight metal deck. It is wiser not to do so in wet, cold areas of the country.

10.1.3 Asphalt coverings

Asphalt is an inherently durable material but can crack if subjected to a sudden stress, particularly at low temperatures. The most common causes of cracking are movement of the deck upon which the asphalt has been laid and differential movement between the deck and features such as skirtings, parapets, verge trims and flashings. Thermal and moisture movements of materials have been considered already and individual values need not be reiterated. It is relevant to note, however, that a flat roof is very exposed and subjected to wide variations in temperature. Asphalt has a high coefficient of thermal expansion and, being black, is likely to be subjected to large temperature changes unless specially protected from solar radiation.

It is probably well known that an isolating membrane needs to be interposed between the asphalt and the deck it protects but failures have happened from time to time because of its omission. Mostly, the omission has been partial, a typical case being that where the membrane has been stopped short of the verges of the roof deck and the asphalt taken beyond the membrane and bonded directly to the deck at the verges. It is less well known that cracking of the asphalt may also be induced if a membrane is not also interposed between the asphalt and any paving tiles or sand/cement screed placed on top as a surface finish, or if too thin a membrane, such as polyethylene sheet, has been used. The detailing of asphalt over expansion joints in the roof deck has often been deficient, when a flush finish is required rather than the twin-kerb recommended in BS CP 144 [65]. Failures have been due mainly to separation of the asphalt from the flanges of flush-type proprietary expansion joints. Partly, this has been through disregard of manufacturers' instructions but the placing of asphalt in conjunction with the materials and joint profiles used to form the joint can be difficult.

Poor detailing at parapets and projections through roofs has been a cause of many failures. Extensive cracking has occurred where asphalt has been dressed up parapets to form a skirting and tucked into chases, without allowance in design for the differential movement between deck and parapet. This has been associated, particularly, with timber and wood-wool decks. Many cases of rain penetration behind the asphalt skirting at the parapet junction have been reported, even when cracking has been avoided. Penetration has been due mostly to slumping of the asphalt from the chase into which it was tucked. The main reasons have been that the asphalt taken up the parapet and into the chase has been too thin; the solid angle fillet of mastic asphalt formed at the junction of roof and parapet has been too narrow at the face; the vertical surface has been too smooth and/or wet to allow the asphalt to adhere properly; and the chase, itself, has been of insufficient size. The lack of a good solar-protective finish will contribute to sagging and blistering. When the horizontal chase is too small, it gives little support to the asphalt where the latter is tucked in and leaves insufficient room for the cement/sand pointing which is required above the asphalt. If the asphalt slumps and shrinks away, rain will penetrate eventually behind the skirting and into the roof deck. The finishing of flat roofs at the edges, using aluminium trim into which the asphalt is dressed has, on the whole, not been very successful and cracking has occurred, particularly at changes in direction of the trim. This, too, is basically because of differential thermal movement, leading to fatigue.

Severe cracking of asphalt has been induced when ordinary emulsion paints have been used as reflective treatments to reduce the absorption

Fig. 10.2 Blistering of asphalt.

of solar heat: such paints can cause deep cracking. Asphalt may also crack if overheated during laying, though this is not usually of major significance unless heating has been grossly excessive. Surface crazing is fairly common and many cases occurred in the fairly hot summer of 1975, with crack widths up to 6 mm. Crazing is due, in general, to failure to sand-rub the asphalt with a wood float. Sand-rubbing not only smooths out imperfections in the final surface but it helps to absorb bitumen brought to the surface during laying of the asphalt. Crazing can be due, also, to the lack of a good surface reflective treatment and to ponding of water on the roof which can lead to large temperature gradients and, thus, to strain, between dry and wet areas. Surface crazing, however, is seldom of great importance and does not lead directly to rain penetration though it may reduce the total effective life of the asphalt.

Asphalt may blister through pressure generated by water vapour, particularly in hot weather, when the pressure can be high. Moreover, in hot weather, the asphalt will be softer and less able to resist such pressures than it would in cold weather. When the roof is of concrete and, particularly, when a lightweight concrete screed has been used, the source of the moisture is mostly the water used for construction. This can be trapped when the asphalt is laid and escape will then be slow: rain may also be trapped. Water may collect, too, as a result of interstitial condensation and may be sufficient in quantity to cause blistering, though this is more likely with built-up felt roofing than with asphalt. Blisters in asphalt can be a sign of omission of, or gaps in, the isolating membrane. Blisters may be of many sizes: diameters of 150 mm are not uncommon. Because the surface area of the blister is greater than that of the corresponding asphalt before blistering, it follows that the asphalt will be thinner than normal over the blister and it may become sufficiently thin to split. It will then, of course, not prevent rain from entering. Many blisters, however, do not split, and do not allow rain penetration, though they will be prone to damage by any foot traffic on the roof and may enhance local ponding and, thus, general deterioration.

It is, of course, necessary that the deck to which it is applied remains sound and provides a stable support to the asphalt. This has not always been the case, particularly when asphalt has been applied to wood-wool slabs. Failures have been caused by the seriously different levels of adjacent slabs and by excessive gaps between slabs. Differences in levels of as much as 18 mm, and gaps of 13 mm, can occur. These uneven surfaces have resulted in the asphalt also being uneven and forming into ridges between slabs leading, in turn, to ponding and cracking. Once wood-wool slabs under asphalt become wet through leakage, the moisture does not escape readily and persistent dampness causes decay of the slab, general loss in strength and disintegration. This leads, in turn, to further cracking and disintegration of the asphalt. A further disadvantage of wood-wool is its high moisture movement. Failures have also been caused by the distortion of some forms of chipboard. Chipboards based on urea-formaldehyde as the binder are highly susceptible to the ingress of water from whatever reason and, when they become wet for any reasonable length of time, weaken and distort. This leads to distortion and failure of the asphalt. Inadequate support of wood-wool slabs, chipboard and strawboard leads to sagging, loss of support for the asphalt and its eventual fracture. Boarded timber decks tend to warp and joints to open, and these have provided an unsuitable substrate for asphalt though, given a high standard of site supervision and construction, satisfactory roofs have been achieved.

Asphalt does not take kindly to damage typically caused by careless handling of scaffolding, the use of concrete mixing plant, bricks and tiles thrown down upon it, spilt paint and solvents, and the temporary storage of heavy equipment which can cause local deformation and indentation. Damage has also been caused by the movement of metal fixings, such as cradle bolts, poorly fixed to the structural deck beneath.

There is a number of actions needed to avoid these failures of splitting, or disintegration, of asphalt, and of leakage. It is first necessary to ensure that differential movement between asphalt and the roof deck and contiguous features, for example, parapets, is minimized. A black sheathing felt, type 4A (1) of BS 747 should be used as the isolating membrane between asphalt and deck. Where the deck and any parapet are likely to experience markedly different thermal or other movement, it is better to apply the asphalt to a free-standing kerb, fixed

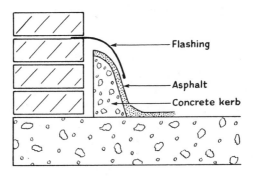

Fig. 10.3 Separation between roof deck and parapet. Free-standing kerb needed when deck and parapet are likely to be subject to markedly different movements.

to the deck, with a clearance between the kerb and the parapet to permit movement. The gap between the kerb and the parapet should be covered with a metal flashing (Fig. 10.3). Where the parapet and deck are not subject to widely differing movement, the asphalt should be tucked into a chase in the parapet. The chase should be not less than 25 mm × 25 mm in cross-section, the exposed part of the asphalt should be splayed to shed rainwater and cement/sand mortar used to point between the top of the asphalt and the underside of the chase. The asphalt will need to be applied in two coats to a total thickness of not less than 13 mm and the solid angle fillet of asphalt formed at the junction of parapet and deck should be not less than 50 mm wide on its face. The height to which the asphalt is taken up the parapet should be at least 150 mm. Smooth and wet surfaces must be avoided and keying to concrete parapets may be necessary, particularly if the concrete was placed against metal or plywood shuttering. Expansion joints are required in the asphalt to coincide with expansion joints in the structure and should be of the twin-kerb variety, if at all possible, for flush expansion joints are very difficult to make waterproof.

If wood-wool slabs are to be used as the deck, it is essential that they are firmly fixed to their supports and are covered by a screed at least 25 mm thick. Adjacent slabs should vary by less than 3 mm in level and should be jointed firmly and taped with scrim. Chipboard should not be used for flat roof decking where occupancy conditions are likely to give rise to high humidity, for example, in laundries. It should be selected and applied as recommended in BS 5669 [66].

The safest solar reflective treatments are mineral aggregates, light in colour and set in a bituminous compound; tiles of asbestos cement similarly bedded; or a sand-and-cement screed cut into paving squares, provided that this is separated from the asphalt by a layer of building paper or similar membrane. Some bituminous aluminium paints may be satisfactory but manufacturers' advice should be sought and followed closely.

Major cracking of asphalt through roof movement implies that separation between the two has not been effected by the proper use of an isolating membrane, and complete replacement may prove necessary. It may be possible, however, if cracking is along one or two lines, to provide movement joints of the upstand variety at these points. Splitting in the region of the angle fillet at parapets will need to be remedied by removal of the asphalt over the full vertical height of the skirting, which should be at least 150 mm, and removal horizontally for about the same distance. A free-standing kerb will need to be provided, with a 13 mm gap between it and the face of the parapet, and detailed as recommended in BS CP 144. Where rain penetration has occurred behind the asphalt at skirtings through inadequacies in the chase, it will be necessary to remove the pointing and the asphalt for a distance of some 60 mm below the chase and to reform the latter, apply fresh asphalt and repoint. Asphalt which has cracked through the application of a harmful reflective paint is likely to need complete replacement. Blisters on asphalt, which have not split, may be left but inspected periodically: they are not necessarily important. However, those which have split will need to be completely opened but can then be patched locally. The removal of entrapped water can be accelerated by dewatering, using a suction pump. This is likely to involve removal and subsequent replacement of the asphalt.

10.1.4 Built-up bituminous felt roofing

Common defects in built-up felt roofing are splitting, blistering, ridging and rippling, local embrittlement and pimpling, and loss of grit. Splitting is caused mainly by excessive differential movement between the felt and the substrate to which it is attached. Felts may be based on organic, asbestos or glass fibres but none can be stretched, without splitting, by more than around 5%, and less if the felt is aged. Differential movement which could cause such an extension occurs commonly. It is because of this that the lowest layer of the usual three-layer felt system is recommended to be only partially bonded to substrates likely to impose undue stress on the felt system. These include concrete and screeded surfaces, screeded wood-wool slabs, particle boards and laminated boards. Extensive movement of the substrate has been due either to its drying shrinkage or through poor fixing to primary supports, such as timber or steel joists. This has been particularly the case with chipboard and with wood-wool. At one time it was believed that felt applied to expanded polystyrene boards should be fully bonded to them for, as with cork and fibreboard, no major differential movement was expected and, hence, no need for partial bonding of the lowest layer of felt. However, many cases of splitting of felt occurred and this is now thought due, not to movement of the polystyrene, but to excessive thermal movement of the felt, caused by the high degree of insulation provided by the polystyrene. This will lead, both annually and daily, to wide variations in the temperature of the felt. The repeated and excessive thermal cycling of the felt has led to its fatigue failure. Some failures, when expanded polystyrene boards have been the substrate, have also been caused by partial melting of the boards through the direct application of them of hot bitumen, used for bonding the felt. This has led to depressions and to poor support for the felt where it bridges over them. Splitting commonly takes place at joints and this can happen even when the felt is only partially bonded, if the width left unbonded is too narrow to take the strain imposed by the joint movement. As with mastic asphalt, built-up felt roofing will split if the substrate fails to provide an adequate base, through excessive distortion caused by lack of strength. Both wood-wool slabs and urea-formaldehyde chipboard are likely to lose strength if they become, and stay, wet, as might happen by leakage of the felt covering, by condensation or by entrapped moisture.

Blisters occur far more readily in built-up felt roofing than they do in mastic asphalt. They form either between layers of felt, commonly

Fig. 10.4 Blistering and ridging of bitumen felt.

under the top layer, or between the felt and the deck or the insulation, if the latter is used above the deck. In both cases, they are due to the expansion of entrapped air or moisture by solar heating. However, blisters should not be taken necessarily as a sign of failure. They are common in practice and often have little significance though, like blisters on asphalt, they present a thinner surface and one, therefore, more prone to damage. Long undulations or ridges may form in felt to give a rippled appearance. In part, this may be due to storage of felts flat, instead of upright, which can cause permanent distortion. In use, there are two main causes. If the ridges are roughly parallel and extend for a considerable distance, the cause is probably their coincidence with gaps and unevenness in level between boards or slabs such as insulation boards or wood-wool. These ridges are essentially hard when pressed. Others, more yielding to pressure, are caused by failure to allow the felt to flatten properly before it is fixed, by poor distribution of bitumen compound used for laying, by insufficient pressure during laying and by expansion of the felt. The entrapped air or moisture expands by solar heating, particularly in summer months when the felt is more flexible and heating is greatest. As with ordinary blistering, ridges may have little significance but do tend to increase ponding and need to be avoided. Minor local cracking, pimpling and pitting, with embrittlement of the felt, is commonly seen. These minor defects are caused by the attack of ultra-violet radiation together with atmospheric oxidation of the bitumen. They are, to an extent, a normal consequence of ageing. This ageing is assisted by the loss of reflective grit which, in turn, will be caused by the slow decay of the thin layer of bitumen used to bond the grit to the top layer of felt. In warm, sunny weather, the more volatile fractions in the bitumen expand and lead to a slow crazing and, ultimately, to exposure of the fibres of the felt. Breakdown cannot be wholly avoided but is accelerated by standing puddles, by the wrong choice of reflective treatment or lack of its maintenance, and by the choice of the wrong felt for the top layer. In the early stages, crazing is superficial but, if not attended to, can deepen and lead, eventually, to failure of the felt to achieve its sole purpose of keeping out the rain. The loss of protective grit through wind and rain is inevitable. It is accelerated by poor practice, such as the use of too small a grit size, by application of grit to the felt in wet weather, by the use of bitumen emulsion for binding and by standing water where the roof has an insufficient fall. Loss of grit, apart from exposing the felt to aggressive solar radiation, may also lead to choking of gutters and rainwater outlets which, in turn, results in longer periods during which water lies on the roof. This is particularly so when integral box gutters are used, for these often have only shallow falls because of the limited depth of the roof structure.

Poor detailing at upstands and angles, particularly at parapets, verges and rooflights, is a common cause of leakage of felt-covered roofs. Blistering, sagging and splitting of felt at skirtings and parapets occur for the same reasons as they do with asphalt (see Fig. 10.5). Cracks between skirtings and the main roof covering are common and are due mainly to differential movement, enhanced by ageing of the felt and the lack of a maintained protective finish. Sometimes, failure has been caused by the simple dressing of the felt into a chase and the omission of a separate flashing. At roof edges, good detailing is necessary to resist uplifting by wind. Metal edge trims have been used to give a neat appearance but are incompatible in thermal movement with the felt. Failure has been caused, either by breakdown in the bond between the felt and the trim, or by splitting of the felt at the joints between adjacent lengths of trim.

As already mentioned, the effects of differential movement between

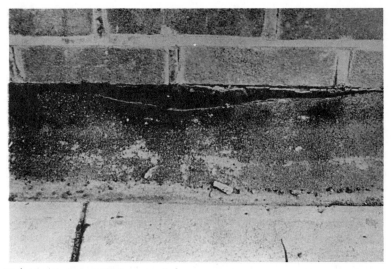

Fig. 10.5 Splitting of felt upstand.

the substrate and the felt are nullified by partial bonding of the lowest layer of felt and this practice needs to be adopted on concrete and screeded roofs, particle board, laminated boards and asbestos decks. It is essential that timber, chipboard and wood-wool slabs are firmly fixed as detailed in BS CP 144 Part 3. Wood-wool is used normally as a lightweight structural deck and to provide some insulation. It is not recommended for use where humidity is likely to be high. If used, the slabs should be pre-felted and left unsealed on the underside. Special methods are necessary when built-up felt roofing is used over expanded polystyrene. The latter should be at least high-duty grade in accordance with the requirements of BS 3837 [67], and should be pre-felted with an underlay complying with BS 747. The underlay felt should be fully bonded to the polystyrene. A vapour barrier of bituminous roofing will be needed between the roof deck and the polystyrene, and a coat of hot bitumen should be applied to the top surface of this vapour barrier, allowed to cool to around 80°C, and the polystyrene boards then placed in position with staggered end joints. Fibre insulating boards, 13 mm thick, should then be coated with hot bitumen and pressed, bitumen-coated side down, onto the polystyrene, with joints staggered to break bond with all joints in the polystyrene. The built-up felt roofing is then fully bonded to the fibreboard by the normal technique.

A warm roof design is preferred and, if dry insulating materials are used, a vapour barrier should be placed between the roof deck and the insulation. No such barrier should be used beneath a wet screed. If a cold roof is used, a ventilated air space will be needed between the insulation and the deck, and a vapour check should be fixed beneath the insulation. All uncompleted work should be covered from the weather. Felts should have side laps of 50 mm and end laps of 75 mm, with the lap joints arranged so that they do not impede roof drainage. Minor movement joints should be provided at lines where movement of the substrate is expected. Felt used for the top layer will need protection from solar radiation. The choice will usually lie between self-finished mineral-surfaced felts of type 2E or 3E of BS 747 or by mineral chippings, 6 to 13 mm in size, secured on site to the felt, using a solution of bitumen in a volatile solvent – not a bitumen emulsion or a hot-applied bitumen. The use of chippings makes any leaks hard to find and does lead to choking of gutters. Some reflective paints may be better but reliable data on performance and durability are lacking, and would-be users should be careful to seek manufacturers' advice. Ordinary emulsion paints should not be used. Where continuous foot traffic is expected, a more durable finish is necessary, as recommended in BS CP 144 Part 3. At upstands, the felt should be taken over an angle fillet and bonded with hot bitumen, to a height of at least 150 mm above the roof. A felt or metal flashing tucked into a chase not less than 25 mm deep will be needed to complete the weathering protection. At roof edges, experience has suggested that welted bitumen-felt trim gives a better protection than metal trim. If aluminium trim is used, it should be in lengths not exceeding 1300 mm, fastened at 300 mm centres, with a 3 mm expansion gap between lengths and covered on its top surface with a high performance felt capping. Where possible, gutters should be clear of the roof rather than integral with it. Bitumen felts are very sensitive to damage and no general building work should be allowed on the finished roof.

The common tendency is to become unduly worried about blisters and ridges, and it is often best to leave well alone but to keep a careful watch for signs of splitting. If repair is decided upon, it is best to do this in warm, dry weather. Blisters and ridges can then be cut out and new patches of felt applied. If blisters and ridges are numerous or extensive, it can be simpler to strip off the felt and re-lay. For minor cracking, it may be sufficient to apply a reflective treatment or a bituminous sealant. Deep cracking or major splitting will require felt to be stripped and fresh felt laid, with an appropriate reflective treatment.

10.1.5 Polymer roofing

In recent years, a range of polymeric materials, including notably poly-isobutylene, butyl rubber, PVC, chlorosulphonated polyethylene and polymer modified bituminous roof membranes reinforced with polyester fleece has become available. Many are designed for use as single-layer systems, stuck to the substrate with special adhesives and the joints between sheets solvent- or heat-welded. Those used in single-layer systems require a high standard of workmanship and most are installed by approved installers. These products are more flexible than the conventional bitumen felts and retain good flexibility after ageing. The main problem has been water ingress through inadequately bonded joints and mechanical damage, causing splits. Such splits are caused relatively easily by stones, nails and tools dropped onto the polymers. Many of these systems have BBA certificates and would seem to have considerable potential. Experience of their use in the UK, however, is still rather limited.

10.1.6 Parapets

Parapets are severely exposed to rain and, also, to wide changes in temperature. They are, therefore, likely to suffer attack by sulphates and by frost. Brick parapets may also be subject to moisture expansion. The causes and symptoms of such attack and movements are considered in Chapter 7. Renderings on parapets are similarly exposed to severe weather conditions. Wherever possible, parapets should be avoided but, if a decision is made to have them, then cavity parapets should be adopted. With solid parapet walls, the problems of preventing rain penetration and the deterioration of any renderings used are such that failure is almost certain (see Fig. 10.6). All bricks used should be low in sulphates. In cavity parapets, failure has been promoted frequently by the lack of an effective coping and by ineffective DPCs in the parapet. The coping system needs to prevent the downward penetration of rain, it needs to throw any rain clear of the face of the parapet wall and the coping material itself must be of high durability to withstand the severe exposure. Brick copings are vulnerable to water penetration at the mortar joints and do not provide an angled weathering surface. In many cases, the bricks used have not had adequate resistance to either frost or sulphate attack and have disintegrated. In other cases, the lack of a DPC immediately under the coping, which is necessary unless the coping itself acts as a DPC (for example, when it is of metal or plastics), has led

Fig. 10.6 Rain penetration through parapet wall.

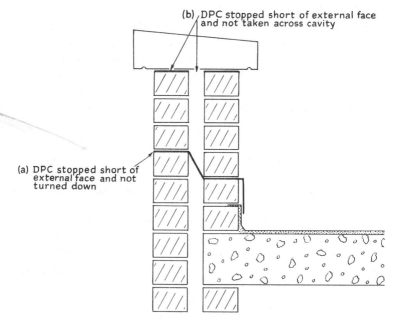

(b) DPC stopped short of external face and not taken across cavity

(a) DPC stopped short of external face and not turned down

Fig. 10.7 Common faults in parapet details. (a) Brickwork becomes wetter and more susceptible to frost and sulphate attack. (b) Rain can run down underside of cavity tray and into roof.

to an unduly wet wall which is more susceptible to attack. A common fault with coping DPCs has been to stop the DPC short of the external face. Water which has passed through the coping, usually through the joints, can then pass into the brickwork below. There have been difficulties, too, when the DPC has been taken under the inner and outer leaves only of a cavity parapet wall and not right across the cavity as well. Rain can then still penetrate through the joints in the coping and drip down into the cavity, ultimately falling onto the brickwork below the DPC level. Unsupported DPCs taken right across the cavity are prone to damage and can sag, presenting some likelihood of separation at lap joints.

With cavity parapets, the coping should not be of brick. Units conforming to BS 5642 [68] would be satisfactory. The coping should be weathered and throated, and its face should project at least 40 mm away from the face of the wall or rendering. The outer edge of the throating should be at least 25 mm away. An effective DPC should span right across the wall, immediately below the coping, should be well bedded between wet mortar and should be supported over the cavity portion. If parapet walls are kept low, it may be possible to take the waterproof covering to the roof up the wall and under the DPC, to form a continuous covering. Parapet walls of a height where this is not possible, because of limitations on the height to which the roof covering can be dressed safely, need a further DPC or cavity tray at the base of the parapet, above the point at which the roof covering is tucked into the parapet. If the DPC is missing or ineffective, then rain clearly can pass behind the edge of the waterproof roof covering which will penetrate only some 25 mm into the wall. Many cases of moisture penetration are known to result from poor detailing of the junction between the DPC at the base of the parapet and the roof covering. A common error has been to stop the DPC short of the flashing which is commonly used to complete the waterproof detail at the roof junction. The cavity tray is usually arranged to slope towards the roof rather than towards the external face of the parapet wall and this has generally been accepted as the most effective form of construction, for it reduces unsightly staining of the external wall. However, it is most important that, in such a case, the tray is taken right through the external wall and turned down to form a projecting drip. If not, and because parapets are so exposed, there is a risk that rain will be conveyed down on the underside of the tray into the inner leaf and there bypass any flashing and the roof covering, and enter the structural roof. Damage can be caused to the DPC when this is located in the same bed joint as the chase cut to take a skirting. Unless considerable care can be assured, it may be better to place the DPC in the bed joint above and to provide a non-ferrous flashing running from the DPC, down the one course of brickwork and lapping over the bituminous covering. Weepholes are needed to assist drainage and, when the DPC slopes towards the roof, should be formed by leaving open perpends every fourth brick in the course of the inner wall of the parapet, immediately above the DPC.

Cracking and general damage to brick parapet walls by frost and sulphate attack will be recognized easily and will usually necessitate the rebuilding of the wall. Before so doing, it will be useful, however, to see whether it is possible to dispense with the parapet altogether. If not, it will be necessary to use bricks of low sulphate content and high frost resistance, and to keep the parapet as low as possible. Excessive moisture movement will result in over-sailing of the DPC at roof level. It may be feasible to rebuild only part of the wall and it will be an advantage to re-use the bricks, which will not display any further appreciable moisture movement. Thermal and moisture movement, frost and sulphate attack may also lead to open joints in coping units and to their distortion. In such cases, the joints will need to be raked out and re-pointed, and the coping replaced with appropriate expansion joints at maximum intervals of 9 metres. Failure of the DPC/roof covering will show as dampness on the upper parts of the external walls and careful examination of the whole parapet system will be necessary to determine the cause. It is usually the DPC system at roof level which is at fault, through splitting or lack of overlap between the DPC and any flashing used. It may be possible to repair the system or to insert new flashings but complete rebuilding may prove necessary. In all operations, care needs to be taken to avoid damage to the roof covering and to provide temporary protection so that, during repairs, rain is prevented from getting under the waterproof covering.

10.2 PITCHED ROOFS

Few problems arise with pitched roofs but, in recent years, some troubles have been experienced with overall instability of trussed rafter roofs. These are now used in probably more than 80% of new domestic dwellings. Trussed rafters are lightweight truss units, generally spaced at intervals of 600 mm and made from timber members of uniform thickness fastened together in one plane, mostly by metal plates, though plywood connectors may also be used. Since their introduction into the UK in the mid-1960s, most have performed successfully, but there have

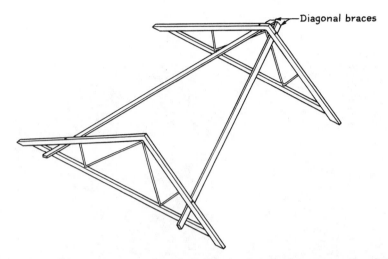

Fig. 10.8 Diagonal braces under rafters, running from heel to apex, are needed for stability.

been cases where trusses have collapsed or moved laterally whilst remaining parallel to one another – the domino effect [69]. This has been caused by the omission of diagonal bracing and the fault is manifested initially by displacement of the roofing tiles. If not corrected, the failure can lead to rain penetration, buckling of rafters and cracking of gable walls. In some cases, failure of the whole structure has occurred through inadequacy of the roof/wall design. The bracing provided for stability of the rafter system, as such, was not adequate, nor designed, to provide stability against wind loading on gable walls. Often, the need for straps for lateral restraint between wall and roof has not been appreciated. The down-turned end of the strap is intended to bear tightly against the external face of the internal leaf, and the body of the strap itself to be fixed firmly to the adjacent truss ties. Common faults are that the straps are fixed to the ties before the internal leaf of blocks reaches that level, and the down-turned end is kept clear of the future face of the blockwork to avoid fouling it. It then serves no useful purpose. In other cases, fixing to the tie is poorly done or omitted altogether [70]. Corrosion of galvanized steel fasteners has also been reported, which suggests insufficient long-term durability, though no failures from this cause are known as yet.

The requirements for truss bracing are now included in BS 5268 Part 3 [71]. Installation of proper bracing at the time of construction is simple

Fig. 10.9 How not to store trussed rafters.

and cheap. The removal of tiles, battens and sarking, and the straightening or replacement of trusses which have moved laterally through failure to provide bracing, will cost at least fifty times as much.

Trussed rafters are intended to be used under dry conditions, and need to be protected and handled with care during storage at the fabricators, during transit and on site. The rafters, when placed, should be covered with the minimum of delay. It is common, and bad, practice for these needs to be ignored, which can lead to distortion, damage to joints and corrosion of the metal fasteners.

Condensation can occur in pitched roofs as well as in flat roofs, and complaints have been made of condensed water running down rafters and even dripping through ceilings. There are several reasons for expecting that the incidence of condensation may rise in modern dwellings unless specific steps are taken to avoid it. Firstly, the increase in thermal insulation, with thicknesses of insulant up to 100 mm now recommended, will make roof spaces above the insulation colder and the dew-point more easily reached. Secondly, the placing of insulation in such thicknesses is likely to impede ventilation at the eaves and, if packed in right up to the eaves, will close the gap between the ceiling

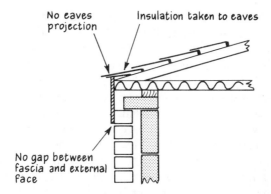

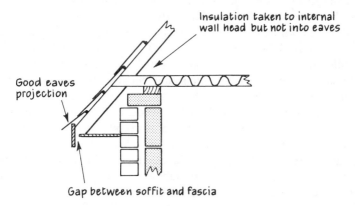

Fig. 10.10 Poor practice: inadequate gap for air to enter behind fascia; insulation obstructs air movement; no eaves projection enhances risk of rain reaching top courses of brickwork if gutters become blocked or defective.

Fig. 10.11 Good practice: gap for external air to enter behind fascia; insulation stopped short of the eaves; good eaves projection gives better protection to the external wall.

and the underside of the roof tiles. This will be so, particularly, when roof pitches are low which, again, is a current trend. Chimneys, too, in many dwellings, are now either omitted or not used, and in such cases, there will be no heat gain from the chimney into the roof space to raise temperatures and so help reduce the risk of condensation. It is now also common practice to use impermeable sarking under the tiles. The main source of entry of water vapour into the roof space is from below and ways of minimizing the generation of moisture in buildings are described in Chapter 4. There are many routes by which moist air, however, can gain access to the loft. The principal ones include the access hatch to the loft, holes around pipes and ceiling light roses, and at the cavity wall heads, which are often only partially closed. Tests have shown [72] that, typically, 50% of the air passing through the ceiling into the roof space does so around the access hatch, 40% through holes around pipes and most of the remainder through ceiling roses. In the loft itself, cold-water tanks may not have well-fitting covers and water can evaporate readily into the loft. This will be especially true when the water in the tank is warmer than the surroundings, as it will be when insulation is not placed under the tank – which, in itself, is good practice in that its omission helps to prevent the water from freezing.

Condensation needs to be prevented as much as possible and a primary reason in roof spaces is to keep the moisture content of the roof timbers at a level low enough to prevent decay. It is also very necessary to prevent the metal plate fasteners in trussed rafter roofs from becoming wet. This is particularly true of the fasteners at the heel joint by the wall plate, which are vital to the structural integrity of the rafter system. This position is the most likely to suffer from lack of ventilation if insulation is packed tightly up to the eaves. It is very necessary to prevent such tight packing and insulation should never be taken right up to the eaves. On the other hand there is the need to ensure that any insulation covers the whole ceiling area or cold spots may occur which could lead to mould growth and to condensation and to corrosion of metal plate fasteners. It is also desirable to close the cavity wall at the wall head by insulating blocks. Figures 10.10 and 10.11 show bad and good practice in roof insulation at the eaves. Positive ventilation of the roof space is essential and the form it should take will be dictated principally by the pitch of the roof. BS 5250 [26] recommends that where the roof pitch is 15° or lower ventilation openings at the eaves should be the equivalent of not less than 25 mm of continuous opening on each side. When the roof is of more than 15° of pitch this figure may be reduced to 10 mm. One way of doing this is to provide appropriate holes in the fascia and soffit boards. Ventilators at the ridge are not an acceptable alternative because they can cause pressure in the roof space to fall under certain wind conditions and this drop in pressure can draw moist air through the ceiling. It can be useful to close the main gaps in the ceiling below by weatherstripping around the access hatch cover, by sealing holes where pipes pass through the ceiling and by ensuring that ceiling roses are screwed tightly to the ceiling. Close covering of cold-water tanks is easily done and useful. However, these measures will only reduce air flow into the roof space; they are unlikely wholly to prevent it. Ventilation of the roof space above insulated ceilings will always be necessary and is, indeed, a requirement of the Building Regulations.

Sheeted industrial roofs are especially likely to suffer from copious condensation promoted by heat radiation to a clear night sky and particularly so on calm, frosty nights. This is an area which calls for expert design and the co-ordinated use of thermal insulation, vapour barriers and positive ventilation.

Tiled pitched roofs may suffer from frost damage but this is not common and damage has been confined mainly to clay tiles used on roofs of shallow pitch. Tiles show signs of spalling on the top surface or may laminate. If the tendency to roofs of lower pitch continues, and if roof spaces become highly insulated, the risk of frost damage to both clay and concrete tiles cannot be ruled out. This is an area where further investigation is needed.

11
Services

Failures in building services seem to lack the journalistic appeal of those affecting flat roofs. Probably, they are fewer in number, anyway, and are characterized more by a falling-off in performance than by total failure. However, some problems have been significant and relate either to heating or to plumbing and drainage.

11.1 HEATING INSTALLATIONS

Many of the complaints made about heating installations have been of their cost in operation and the difficulties of using alternative fuels, through inflexibility in building design. Many housing estates, for example, are either wired for electricity or piped for gas but do not enjoy both, and many modern dwellings do not have chimneys and are not able to use solid fuel without alteration. These issues are essentially social and economic, and lie outside the scope of this book. In passing, however, it is worth restating that many occupiers have, as a consequence, chosen to use portable fuels, such as paraffin and liquefied petroleum gas, which generate large quantities of moisture during combustion and so increase the risk of condensation and the problems which stem from it.

There have been technical problems, however, in the post-war years, particularly, those of scale formation in hard-water areas and corrosion.

11.1.1 Scale formation

Hot water pumped through radiators is, probably, the most common form of central heating. The composition of the water supplied is dependent upon the nature of the ground in the area in which the water originates and this will vary between individual water-supplying authorities. Rain absorbs carbon dioxide from the air to form carbonic acid and this, draining through calcium-bearing soils, will form calcium bicarbonate. When heated, this decomposes to calcium carbonate which is, for practical purposes, insoluble. Because calcium bicarbonate is so changed when water is boiled, it is said to cause 'temporary' hardness. The equivalent amount of calcium carbonate in water is used as a classification of its hardness. Water draining through chalk may have more than 350 parts per million and is then classified as very hard. That derived from peaty areas is likely to have less than 50 ppm and is classified as soft. 'Permanent' hardness is not removed by boiling and is caused mostly by the presence of calcium and magnesium sulphates. In a hot-water system, calcium carbonate will only continue to be deposited, and to build up to form a harmful scale, if the water is constantly replaced by further bicarbonate-bearing water. This will happen in a direct hot-water system in which the water heated by the boiler is drawn off and replenished by a new supply of cold water. Scale formed in this way in direct hot-water systems reduces the effectiveness of the boiler and, eventually, causes the entire blockage of water-ways. The rate of formation of scale increases with increasing temperature and with increasing degree of hardness. In recent years, many direct systems have been badly affected by blockage. Scale formation through permanent hardness can also cause blockage but the rate of formation is very slow. In part, failure has been increased by the introduction of smaller boilers with narrower water-ways.

Problems of blockage and furring of pipes, through scale formation, can be effectively prevented by adopting an indirect hot-water system. This is a system in which water heated by the boiler is not drawn off but is used to raise the temperature of water in an indirect storage cylinder. This it does by circulating through a heat exchanger within the cylinder so that it does not mix with the stored water. In consequence, and except for any small amounts of water lost by expansion or leakage, the same hot water circulates in the primary circuit and, after the initial deposition, no further scale can form. Indirect systems are the most commonly used today and blockage by scale formation should be a declining problem.

11.1.2 Corrosion

Separate from the malfunctioning of central-heating systems through scale formation can be that due to corrosion. Severe problems have been encountered in indirect heating systems, where the same hot water is pumped continuously round a closed circuit. The heating circuits commonly involved have used cast-iron boilers, copper tubing and pressed steel radiators and most problems have occurred in centralized systems heating a number of dwellings at a time. In an indirect system, the initial water quite quickly loses dissolved oxygen, and iron in the circuit then reacts with this water to form soluble iron products. This, in itself, is not particularly troublesome and an equilibrium is soon reached. Other factors, however, may cause the iron compounds to break down with the formation and release of hydrogen. Further iron compounds may then form and be broken down, until a sludge of less soluble iron compounds is formed. This and the entrapped hydrogen may together stop the circulation of water within the system and also prevent the proper functioning of valves. The causes of the generation of hydrogen and the break-down of iron compounds into a less soluble form seem not to be understood fully, but copper, dissolved in small amounts in the water, is thought to be a contributory factor. When blockage has occurred, flushing out and chemical treatment to remove the sludge are the likely actions necessary but can be expensive.

Bimetallic corrosion, which is considered in some detail in Chapter 3, can be troublesome when oxygen is present in the water system but, with indirect systems, the oxygen content is generally low. However, it may be introduced through faulty design, for example, if the circulating pump operates at too high a pressure and if air can enter the cold-feed pipe to the boiler. This can happen if the water level in the expansion or 'header' tank ever falls below the level of that pipe. A common cause of pitting corrosion of pressed steel radiators has been the continual entry of air into the system promoting copper/steel bimetallic corrosion. Failures have also been associated with the formation of nitrogen or hydrogen sulphide by forms of bacteria. The frequent formation of air-locks, in addition to the initial air-locking which is not unusual in a new system, is an early sign of the likelihood of corrosion and blockage. If a check of the system shows no reason for leakage and entry of air, and the water level in the header tank is not too low and does not overflow either, then it is probable that a corrosion inhibitor containing a biocide, placed in the primary water supply to the boiler, will prove effective in reducing corrosion. However, it is essential that the local water authority is consulted before using an inhibitor because not all inhibitor/biocides are harmless to health, and in no circumstances must they be allowed to enter, or contaminate, the drawn-off water supply. They are for use only in closed primary circuits.

11.2 CHIMNEYS AND FLUES

Some years ago, extensive failures occurred in chimney stacks, particularly those serving slow-combustion solid-fuel boilers and closed stoves. Water vapour is always present in flue gases, which may also contain sulphur oxides and tar acids. As the flue gases pass up the chimney and reach the exposed part of the stack, the dew-point is commonly reached and condensation takes place. The condensate migrates into the walls of the chimney stack, and may deposit tarry residues and, also, set up sulphate attack on the mortar, through the sulphur gases dissolved in it. Sometimes, additional sulphates contained in the bricks assist the attack. As already described, sulphate attack causes the mortar to expand, and this leads to distortion of the stack, which can be extensive, particularly on rendered stacks (see Fig. 11.2). Stacks often bend in one direction and this may be due to preferential wetting of one side of the stack by rain predominantly from one direction. The condensate often contains salts, which absorb water vapour from the air and serve to keep affected parts damp. Both external brickwork and internal plaster on the chimney breast are likely to be discoloured by the tarry compounds, particularly at ceiling level: dampness will be visible in wet and humid weather. (If dampness, however, only corresponds with periods of rain, it is probable that the flashings and DPC in the stack are missing or faulty.)

For some time now, the Building Regulations have required all new chimneys to be lined with materials suitably resistant to acids, and impermeable to liquids and vapours, and this problem should be one of the past, and of passing, rather than direct, interest. Repair to existing damaged chimneys will often require rebuilding of the stack and relining but, if distortion and damage are slight, it may prove possible to insert a suitable lining without dismantling.

11.3 PLUMBING AND DRAINAGE

The soils through which cold-water supply services pass can vary considerably, from comparatively non-aggressive light sand and chalk to heavy clays containing organic matter suitable for the development of bacteria which can assist corrosion. Some sulphate- and chloride-bearing soils can be very aggressive. Ground to a depth of up to 1200 mm is of main interest and here the natural soil may be contaminated, anyway, by a wide variety of builders' rubble. When corrosion or other faults do occur in buried pipework, repair tends to be costly. Protective tapes and wrappings are called for in many areas where galvanized steel is used. Protective tapes are usually recommended to be wrapped spirally around the pipe, with at least a 50% overlap, giving a double layer of tape. Wrapping is a time-consuming task, however, and one which requires good supervision if gaps are not to be left which will

Fig. 11.1 Deterioration of brickwork typical of sulphate attack.

Fig. 11.2 Expansion of flue and the start of sulphate attack on the rendering.

present an area for concentrated attack. Some tapes, too, can be degraded by bacterial activity and fail unless they, in turn, are protected by an outer, inert plastics tape, also wrapped spirally to give a double thickness.

Protective tapes have been damaged, and made ineffective, by insufficient care in back-filling the trench, in particular, by sharp-edged builders' rubble able to cut the tape. Trenches need to be backfilled, to a depth of around 300 mm, with selected fine material, such as sand, hand-packed around the pipe. Unfortunately, authoritative data on the best types of protective tape to use are lacking, partly due, no doubt, to the wide variety of corrosive and bacterial activity likely to be encountered. It may be useful to undertake a soil survey or to refer to existing surveys to determine the likely corrosivity and probability of bacterial activity. Lead is prohibited for all new work in England, Wales and Scotland. Copper may be used but this, too, can be readily attacked by rubble and will often need a protective coating. Polyethylene and unplasticized PVC are used also for cold-water supply and do not corrode, though they can be damaged by poor back-filling and by careless handling at low temperatures. They do not require taping, and unplasticized PVC pipes may well be a preferred solution in active soils. (Pipes made from plastics are not suitable for hot water, and may soften and sag if placed near to hot pipes and where subject to undue temperature rises, such as in airing-cupboards.)

Under damp conditions, pipes passing into the structure through brick, concrete and plaster can be corroded. Pitting corrosion of stainless steel can occur when the unsheathed steel passes through breeze blocks, possibly caused by the presence of small amounts of chlorides. All metal pipes passing through these building materials should be suitably coated, taped or sheathed: it is safest to assume that, sometime or other, they will become damp. However, failures are few and the problem is not a significant one nationally.

Internally, corrosion in plumbing services is caused principally by the use of mixed metals in the system. Of the mixed metals used, the combination which has given rise to most failures has been copper/galvanized steel. It is the galvanized steel which will corrode. The corrosion rate increases if the water contains much oxygen or chloride ions. The chief risk is when water flow is from the copper part of the system to the galvanized steel part and when new copper components are introduced into an old system using a galvanized steel tank. However, if flow is from a galvanized steel tank to the copper, and if copper in solution in the hot-water part of the system does not find its way back to the galvanized tank, for example, via the expansion pipe, the risk of corrosion is greatly reduced. The use of galvanized steel and copper together in a hot-water system will almost certainly lead to rapid failure of the former. Such a combination is unlikely in new buildings but many older ones contain galvanized steel pipes used before copper came into favour. The introduction of a new copper cylinder in such cases can cause rapid failure. Stainless steel is now moving slowly into use but acceptance is retarded by difficulties in jointing, caused, in part, by its low thermal conductivity which requires different techniques of soldering. The fluxes commonly used for soldering copper pipes are based on zinc chloride and were used with stainless steel when the latter was first introduced. Severe pitting corrosion has occurred when such fluxes have been used excessively and when a long time has elapsed between making the joint and passing water through the pipe. Chloride residues need to be completely removed from stainless steel after application, otherwise pitting corrosion occurs. The need for such care, and the evidence of severe corrosion in its absence, has retarded acceptance of stainless steel. Fluxes based on phosphoric acid have been introduced, which are believed to be much safer. In the absence of chlorides, stainless steel in plumbing systems should not, itself, corrode and will not cause corrosion of copper.

Minor troubles can affect ball valves though this has been of small significance. Ball valves can stick in the open position, leading to overflowing or flooding if the overflow pipe is blocked or defective. Generally, this is merely a sign of the need for a new washer but the seating may be eroded by high-velocity water discharge. Many older floats, formed of two copper hemispheres soldered together, can corrode, become perforated and partly fill with water, which then prevents the closure of the ball valve. They may drop away from the float arm, leading to rapid overflow. Mostly, however, plastics floats are now used and these do not corrode and give much better long-term performance. Calcium carbonate deposits in hard water areas forming on the piston of the valve cause this to stick and are a common reason for overflowing.

Finally, a few words on frost damage. Pipes placed above insulation in a roof space will get cold. Lagging will slow down the rate of heat loss but, if heat gain into the roof space either from below or from other pipes is too small, freezing may still occur. Lagging will merely postpone eventual freezing in an unheated building. With the likelihood of increased insulation in roof spaces, the risk of freezing may also increase, unless the pipes are taken under the insulation wherever possible or some other steps are taken to ensure a low heat input to the pipes.

11.4 ELECTRICITY SUPPLY

All electric cables give off heat in use and this is usually dissipated without any difficulty. However, cables can get overheated if placed beneath loft insulation, behind insulated dry lining or if placed in a position where the temperature of the surrounding environment is high. The insulation of the cable will then be damaged and there can be a risk of a short-circuit and, possibly, fire. Cable in power circuits is more at risk than cable in lighting circuits for the former is more likely to be loaded near to full capacity. Electric cable should not be covered by thermal insulation nor, ideally, should it be used where ambient temperature regularly exceeds 30°C. Where this is not feasible cables will need to be de-rated in accordance with the regulations prepared by the Institution of Electrical Engineers. It may be worth noting that a cable de-rating factor as low as 0·5 will be needed where cable is insulated on both sides. There can also be an interaction between PVC cables and expanded polystyrene often used for insulation and this can cause degradation of the PVC. The two should be kept apart.

12

Failure patterns
and control

This last chapter is different in nature from those preceding, and, after commenting on failure patterns, suggests reasons why avoidable building failures occur. Such reasons involve consideration of the structure of the industry, the particular problems associated with innovation, and control systems. These essentially 'software' aspects of the avoidance of building failures merit as much, and probably more, attention than the 'hardware' aspects dealt with already. It is, indeed, probable that the greatest scope for improvement in building performance lies more with the former than the latter.

While there is a considerable consensus of opinion on the technical reasons for failure which, it is believed, has been reflected in the preceding chapters, the 'software' aspects are more speculative. The views on these are many and varied. It is hoped that those expressed here will be of some interest and possible value.

12.1 COST AND TYPES OF DEFECT

There are no precise data on the total national cost of repairing avoidable defects. Only in a few major organizations in the public sector are there systems in operation which aim to record defects systematically. The private sector is too diffuse for any worthwhile attempt to have yet been made to collect such data. In the housing side of the private sector, many defects will be rectified, anyway, by uncosted 'do-it-yourself' activity. Typical maintenance budgets in both public and private sectors cover normal maintenance due to fair wear and tear and the cost of cleaning, as well as repair.

In considering costs, problems of definition arise in deciding what is an avoidable defect and whether the estimate of cost should include not only that incurred in rectifying, physically, the defect, but also the consequential costs, such as loss of rent or production and temporary resettlement. Consequential costs may be much greater than direct repair costs and the wide range of guesses at the cost of failures is doubtless partly due to the different boundaries within which the costing assessment is made. Nevertheless the defect reporting systems of the Property Services Agency (PSA) reveal that more than 10% of the Agency's maintenance bill is attributable to preventable defects occurring within about five years of construction [73]. Nationally, the total cost of repair and maintenance is estimated to be in excess of £12 000 million [74]. The implication is that preventable defects are likely to be costing somewhere in the region of £1000 million a year.

There are also no reliable data on the relative proportions, nationally, of different types of defect by number and significance. In 1975, an analysis was made, however, of over 500 investigations undertaken by the Building Research Advisory Service [75]. The report emphasized that the problems were not necessarily representative of the generality of building defects. In the analysis, dampness, whether from rain penetration, condensation, entrapped water or rising damp, accounted for half of all the defects investigated. Condensation was the largest single cause of complaint in council housing and rain penetration in private housing. It is probable that, since the analysis was made, condensation has retained its pre-eminence as a problem in council housing. Condensation, of course, is not necessarily caused by a design or construction failure, though it may be exacerbated by them. Rain penetration, on the other hand, clearly is, and the analysis showed that most cases reported were those of penetration through walls including parapet walls, followed closely by penetration through roofs. Cracking, both structural and superficial, and detachment of finishes, such as rendering, brick slips and tiles, accounted for some two-thirds of all other reported defects. The cost significance of the various defects analysed was not reported.

Analysis of defects starts, generally, with a classification of symptoms, for example, a damp ceiling as a symptom. This is the easy part. A more penetrating diagnosis will then be required to determine the cause of the symptom, for example, split roofing felt. Still further analysis will be needed to determine the cause of the splitting, such as differential movement between felt and substrate. The ultimate cause of the differential movement will lie with one or more of the agencies mentioned in Chapter 2. These agencies are the real physical, chemical and biological causes of failures. The deeper and more efficient the analysis, the greater will be the success of a remedial measure. All too

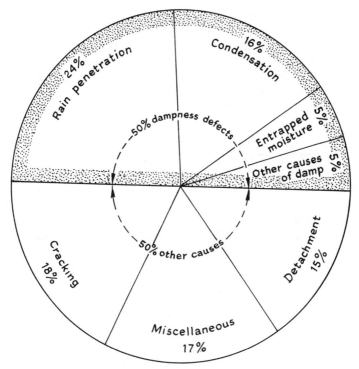

Fig. 12.1 Analysis of defects.

often, the analysis stops prematurely and a facile solution is adopted, which may be unsuccessful or only partially successful. Good design and construction, aimed at preventing premature failure, should start with an understanding of the basic agencies likely to affect the element of structure in question, assess their effects and then design to minimize or prevent them. In general, this synthesis is performed intuitively rather than objectively.

12.2 CAUSES

In the study by the Building Research Advisory Service, the causes of failures were also analysed to indicate whether they were due to faulty design, to poor execution, to the use of poor materials or through unexpected user requirements. Faulty design was taken to include all cases where the failure could reasonably be attributed to a failure to follow established design criteria: 58% of all failures were found to lie in this category. Faulty execution, defects attributed to a failure of the builder to carry out a design satisfactory in itself and properly specified, accounted for 35%. Only in 12% of cases did the materials or components fail to meet their generally accepted performance levels. Some 11% of failures were caused by the users expecting more from the design than the designer anticipated. (There was some overlap between these categories.) A more recent analysis within the defect reporting systems of the PSA already referred to showed that 51% of the causes of defects were due primarily to shortcomings in design, 28% to inadequate workmanship and the remainder were due to other causes including the failure of materials [76]. A recent BRE study into faults in conventional two-storey house building in England found that 50% were attributable to design, 41% to site and only 8% to other causes including materials. The fact that far more failures occurred through inadequacies of design or execution than through faulty materials was a result which is confirmed by most subjective assessments.

12.3 REASONS FOR FAILURE

There seems little doubt that a major reason for failure in construction is the complex structure of the building industry. Despite the calls for closer integration of design and construction, made many times over the years, these two vital roles are still essentially separate. The manufacture of the basic materials and components, too, upon which the industry depends, is dispersed among many firms. These vary in size and type, from large firms, often with their roots in other industries, for example, the chemical industry, to small firms supplying minor components.

The building contractors, in the main, are small, with most employing fewer than ten operatives. There are more than 30 000 general builders and a similar number of specialist contractors, such as painters and plumbers. Some 70% of total output is attributable to contractors employing fewer than twenty people. In recent years, there has been a marked change to extensive subcontracting by the main contractor and a noticeable trend towards labour-only contracting. These structural complexities in this large and diffuse industry are conducive to difficulties in management, in education and in training; in the communication and application of technical information; and in the understanding and acceptance of responsibility. Construction, as a whole, has moved away from being craft-based towards being industrialized. However, it is currently in a difficult in-between state and the clear loss of craft skills has not been replaced by the type of relatively fault-free product typical of truly industrialized processes. The loss of craft skills has been matched at a more professional level by a reduction in the number of those able to comment sensibly on the likely interactive effects of changes in materials, components and procedures. Many of the problems which have beset so-called industrialized building systems have stemmed directly from the general inability in the construction industry to understand these interactions, together with inadequate matching of site skills to the new technology.

It is difficult, clearly, for technical information at the right level to flow steadily and smoothly to the right recipients in such a diverse and changing industry. Shorter lines of communication, which obtain in more coherent and concentrated industries, would do much to improve the flow but there are few signs, as yet, of change in the basic structure of the industry. In some cases, the technical recommendations, although available, are not *readily* available and while they do state 'the truth and nothing but the truth' it is, sometimes, not 'the whole truth'. Thus, some failures, though probably a relatively small portion of the whole, do stem from knowledge which is less than comprehensive. It is probable that most failures, however, could have been prevented by the proper application of existing technical knowledge. Technical information is produced in many forms but its diversity equals, rather than matches, that of the industry. The knowledge within the UK is probably as comprehensive as that in other countries but it is fragmented. Reliable, self-contained and comprehensive guides to specific design and building operations, written at the various levels needed to match the skills and abilities of the constituent parts of the construction industry, are in general not available. However, a recent development has been the setting up at the BRE in 1981 by the Minister of Housing and Construction of a Housing Defects Prevention Unit. This Unit sends to contact points within local authorities short, sharp advice through its well-illustrated Defect Action Sheets aimed to prevent, in housing, the most common defects in design and in site practice. As housing, both in new construction and in repair and maintenance, accounts for about one half of total construction output this initiative is an encouraging one.

In addition to the problems engendered by the structure of the industry and the difficulties it causes for the capture, dissemination and implementation of technical knowledge, are those of innovation, control and product certification. The construction industry has to operate within a complex framework of control, which is not uniform within the UK and within which, in recent years, the final responsibility for failure seems unclear. Moreover, the post-war years have seen rapid innovation in both products and methods. While innovatory pressures have been considerable, the requirement for adequate control and certification of these has lacked 'bite'. Too often within the innovatory decision-making process the views of those with a mature knowledge of building performance have carried less weight than those pursuing technical change.

12.4 PROBLEMS OF INNOVATION

Pressure for innovation may arise through events affecting the nation as a whole – for example, the present need to conserve energy and the consequent legislation which has exerted a pressure for new or improved ways of insulating buildings. Actual or potential scarcity of resources, too, has been a potent influence on innovation. The resources may be of labour and particular skills, or of materials. A related pressure may be the need, assumed or real, for speed. Such pressure usually follows situations of greatest social and economic change, for example wars, and was the main one behind the non-traditional construction methods urged after both the First and Second World Wars, with unhappy results. The detrimental environmental effects of accumulation of waste of other industries, and the costs of its disposal, have also resulted in pressure on the building industry, because of its large market size, to take part of the waste as a raw material. Inflation, too, tends to cause pressure for labour-saving innovation in industries with a high labour content, such as the building industry. These strategic pressures find expression in entrepreneurial innovation where, rightly or wrongly, it is seen as likely to increase profitability or as essential to maintain competitiveness. Quite often, the creative thought and its transformation to a marketed product or method derives from other industries. The chemical industry and, in particular, the plastics part of it, is the most obvious example.

When the general social and economic system remains stable, the pressure for innovation is reduced: where it is unstable, innovation is a natural reaction. Change is not necessarily progress. It may create healthy competition but may also destroy skills which cannot so readily be re-created. Poor innovation can, unfortunately, be self-perpetuating, creating the need for remedial innovation. In contrast, the more radical and better the innovation, the more it helps to restrict the climate for further innovation. It would be interesting to determine, if such a thing were feasible, the relationship in the building industry between numbers of innovations and general quality of building: it might well be inversely proportional.

These pressures do not arise from consumer choice but from industry, commerce and Government. This is, of course, a not unusual state of affairs in other sectors of industry and stems in part from the lack of a suitable mechanism to obtain users' requirements in any meaningful way and the inability of many users, through ignorance, to express valid opinions.

While these pressures for innovation are real and active, there are constraints to *responsible* innovation, several of them peculiar to the building industry. The major constraint lies in its organization. The building industry is not an entity which, by its organizational and financial structure, permits the injection of large resources to encourage responsible innovation and there seem few signs of early change. Innovation, in fact, usually comes from the manufacturer and supplier of materials and components.

Innovation in the building industry is a somewhat random process, most often directed at reducing initial cost, though protective pricing, to provide a margin against uncertainty, may make this aim imperfectly realized. The lowest initial cost concept, a corner-stone of most of the associated advertising, almost always results in a reduction in the quality of materials, services and structural components, or in the amount of usable space in the building. Such innovation slowly tends towards shortening total life and increasing the need for maintenance. This is in opposition to the true needs of conserving the building stock and reducing the need for repair, which is higher in labour content than that for initial building. It is seldom that innovation takes place against a criterion of total costs in use. To be successful commercially, the time scale for innovation, development, production and marketing needs to be as short as possible but this contrasts with the basic development in building which has evolved over a long period. The tendency is to innovate at the trivial end, rather than to consider control of the whole system and its likely future needs. The innovator seldom has a clear grasp of the required future state of the building stock – only a knowledge, usually, of one isolated facet of the state he aims to change. The most common danger is to ignore, or to be unaware of, the often complex interactions which the innovation will alter. The innovation is nearly always concerned with providing an alternative to the primary function of an existing material or system but seldom has regard to important, though often ill-defined, secondary features. A typical paradox, through this lack of total understanding, is the way improvements in the prevention of heat loss may increase the risk of condensation or rain penetration: the solution provided by the innovation is often subtly different from that intended. A further danger is the inability to appreciate, or at least to take account of, erection accuracies and the level of supervision reasonably possible on building sites, and failure to provide margins against slight misuse, both in design and construction.

In the longer term, and at its most extreme, there is a possibility that innovation will make existing knowledge and skills redundant. If the innovation is enduring and valuable, this may be acceptable but, if not, the nation will be the loser. But, perhaps, the biggest danger of all lies in the inadequate mechanism for control of the innovation to safeguard the user.

12.5 TOWARDS BETTER CONTROL

Any change in the basic structure of the building industry is likely to be slow and brought about by economic forces rather than by any conscious desire to reduce the incidence of failures. It seems probable that the continuing fall in the number of building operatives, and the increasing cost of the labour, will comprise an economic force likely to enhance the development of semi-industrialized building techniques, leading to faster erection and requiring less skill. In the long term, this should improve the interface between design and construction, by the encouragement of 'design-and-construct' organizations. The harder economic climate is also likely to reduce the number of small contractors and lead to amalgamations which will, in theory, provide better targets for technical information and for feed-back.

The principal organizations, both public and private, have the ability to reduce the risk of failures by taking formal steps to provide an effective interface between design and maintenance. Maintenance organizations are the repository of much knowledge of the nature of failures and their cause, and such knowledge fed back into design would be of great benefit. A dialogue at the design stage could do much to make designers more conscious of the implications of their design decisions and the ability of contractors to fulfil them safely, and should lead, if such a dialogue is conducted with tact, to alternative and less sensitive design decisions. The large organizations might have the resources to adopt the formal collection and feed-back of information relating to failures, but even the more modestly sized organizations could, it is suggested, establish such a dialogue. The PSA design/

maintenance liaison (DML) system was set up as a vital link to improve the quality of building and to eliminate premature maintenance costs. It has clearly proved its worth within the public sector. There is a general need throughout the construction industry to discourage designers from using novel solutions to problems where standard solutions are already available.

There is still a need to improve the control of innovation. The British Board of Agrément's function is to assess, test and, where appropriate, issue certificates in respect of materials, products, systems and techniques used in the construction industry, particularly those of an innovative nature, in order to facilitate their ready acceptance and their safe and effective use. Agrément certificates provide the Board's opinion of the fitness for specified purposes of materials, products, systems and techniques taking into account the context in which they are to be used. While a large number of certificates have been issued there is no compulsion for an innovation to be referred to the BBA. The process of designing and building is often achieved with little involvement of the professions within the industry and as a result products and processes may be used which are unsuitable. If BBA certificates were demanded, if only for certain categories of materials and components, then manufacturers would find it more worth while to pay for the evaluative testing and to modify the innovation if necessary. As it is, a manu-facturer accepts the need to obtain a certificate only if he sees this as of commercial benefit. Building Regulations provide some brake on innovation. They are, however, made for specific purposes of health and safety, energy conservation and the welfare and convenience of disabled people. They are not concerned with other aspects of per-formance such as serviceability.

Innovations should be examined in relation to total-life costs, with particular emphasis on improving durability and reducing the need for maintenance. There is a need to move away from the current concept of trying to encourage innovation by paring safety margins and quality, and to move towards innovation aimed at reducing obsolescence and providing greater adaptability. The risks of innovation should be more truly balanced against the claims by in-depth studies of the whole system. As a start, priority should be given to examination of innova-tions affecting the structure of buildings. A second order of priority should attach to those innovations which, while not directly affecting safety, may, if they fail, cause big expenditure. Studies of innovation by the innovator, and any control system, should include full consideration of the life aimed at; the positive evidence for the likelihood of its achievement; the likely extreme and normal environmental conditions; the probable loads; the factors which might cause failure; the probability of occurrence of such factors; the consequence of failure; the warnings needed for the designer and builder; the quality of control desirable; the strength of control which can be exercised by the innovator over the use of the product; the skills needed in construction likely to be available; the margin of permissible error and the type, and ease of undertaking, of any future maintenance.

In support of responsible innovation, there is a need for greater knowledge of how buildings and their elements actually behave, as opposed to the assumptions so often made and enshrined in British Standards. This implies major planned and purposeful site surveys. Associated with these should be an accelerated development of performance requirements and their related evaluation procedures.

In recent years there have been significant developments in the field of quality assurance, in particular the Government initiatives set out in a White Paper [77] and BS 5750 'Quality Systems', the latter being concerned with the quality of entire management systems. There are now BSI Registered Firms – firms certified by BSI as having a quality management system in accordance with BS 5750. Some of these firms are also holders of BBA certificates. Failures are not often due to the poor quality of materials and components. Nevertheless, the appraisal and control of the quality of these is more marked here than it is in design offices and on sites where it is most needed. Doubtless, this is because it is easier to do so. However, steps are needed to improve significantly the quality control of designs and specifications. In design offices, much of the detailing and specification is the task of the more junior and less experienced designers, who have little knowledge of what is practicable on site, and the abuses to which materials and components are subjected. More effort is needed to develop standard solutions and details. In general, the typical site operative will do what is quickest and easiest. If details and whole designs can be devised so that what is quickest and easiest is also sound and durable, then a big advance would be made. Design will benefit considerably if design offices are organized to ensure that the experience and knowledge of senior designers is brought to bear, both in this vital area and in the independent checking of detail.

The need for improved quality of site operations is clear but the way forward, obscure. Sooner or later, the main contractor will need to regain control over subcontract labour, and site supervision will need to be more concerned with the quality of the finished product and less with accepting shoddy building in the interests of more harmonious site relations. This control may, in part, be brought about by financial penalty or by some return to directly employed tradesmen. It may also be assisted, particularly in housing, by the development of industrialized systems which require less skill and less labour for erection. Any such systems will need to be designed with much greater care than in the past. Ideally, the erection of the individual components should only be possible in one way, and that one should be the easiest. Wherever possible the design of any building should facilitate ease of construction.

Improvements are needed in the dissemination of technical information at all levels but, particularly, between the designer and contractor. Design information should be complete and the builder should not need to spend his time in ascertaining the designer's real intentions or in devising his own, often inadequate, design solutions to problems. Drawings and specifications should be complete and the information on, or in, them easily found by using some structured system. As already mentioned, there is not so much a lack of information as a poor match between the level of information and the recipient. There is a particular need for more detailed, specific, self-contained and readable guidance to be given on the design of individual building elements, and for parallel guidance to builders and operatives on site to meet the design intentions. British Standards could provide the starting-point for such guides but are not, as they stand, appropriate.

There is also a need for designers to be aware, in a far more specific way, of the likelihood and cause of failures. Their education and training needs to encompass the most common types of failure and their cause. Designers also need, during their training, guidance to the sources of reliable information on design principles and solutions, to avoid failures. It is hoped that this book will go some small way in assisting such education and training.

References

ABBREVIATIONS

BRE   Building Research Establishment
BSI   British Standards Institution

NOTES

BRE   Digests are published by HMSO, London
BSI   Standards are published by the British Standards Institution, 2 Park St, London

1. BRE Digest 299, *Dry rot: its recognition and control.*
2. Department of the Environment and the Welsh Office (1985), The Manual and Approved Documents to the Building Regulations (England and Wales), HMSO, London.
3. BSI (1973 and 1974) *Asbestos-cement slate and sheets*, BS 690 (Part 3, *Corrugated sheets*, 1973), (Part 4, *Slates*, 1974).
4. BSI (1985) *Mastic asphalt for building (natural rock asphalt aggregate)*, BS 6577.
5. BSI (1974) *Clay bricks and blocks*, BS 3921.
6. BSI (1978) *Calcium silicate (sandlime and flintlime) bricks*, BS 187.
7. BSI (1981) *Precast concrete masonry units*, BS 6073, Part 1.
8. BSI (1971) *Concrete roofing tiles and fittings*, BS 473.
9. BSI (1979) *Specification for clay plain roofing tiles and fittings*, BS 402.
10. BRE Digest 35, *Shrinkage of natural aggregates in concrete.*
11. BRE Digest 258, *Alkali aggregate reactions in concrete.*
12. BSI (1980) *Sulphate-resisting Portland cement*, BS 4027.
13. BSI (1985) *The structural use of concrete, code of practice for design and construction*, BS 8110.
14. Department of Education and Science (1973) Report on the Collapse of the Roof of the Assembly Hall of the Camden School for Girls, HMSO, London.

15. Bate, S. C. C. (1974) Report on the Failure of Roof Beams at Sir John Cass's Foundation and Red Coat Church of England Secondary School, Stepney, BRE Current Paper 58/74, The Station, Garston, Herts.

16. BSI (1971) *Copper tubes for water, gas and sanitation*, BS 2871, Part 1.

17. BRE Digest 269, *The selection of natural building stone*.

18. BRE Digest 177, *Decay and conservation of stone masonry*.

19. BSI (1971) *Roofing slates*, BS 680.

20. BSI (1976) *Code of Practice for Stone Masonry*, BS 5390.

21. Princes Risborough Research Laboratory (1982) Technical Note 38, The Movement of Timbers, The Laboratory, Princes Risborough.

22. BSI (1984) *Code of practice for the structural use of timber; permissible stress design, materials and workmanship*, BS 5268, Part 2.

23. Lacy, R. E. (1976) *Driving Rain Index*, HMSO, London.

24. BSI (1984) *Methods for assessing exposure to wind-driven rain*, BS DD 93.

25. BRE Digest 290, *Loads on roofs from snow drifting against vertical obstructions and in valleys.*

26. BSI (1975) *Code of basic data for the design of buildings: the control of condensation in dwellings*, BS 5250.

27. BSI (1981) *Site investigations*, BS 5930.

28. Ward, W. H. (1948) The Effect of Vegetation on the Settlement of Structures, Proceedings Conference on Biology and Civil Engineering, ICE, London.

29. BSI (1972) *Foundations*, CP 2004.

30. BSI (1972) *Foundations and substructures for non-industrial buildings of not more than four storeys*, CP 101.

31. National House-Building Council (1986) *Registered House-Builders Foundations Manual: Preventing Foundation Failures in New Dwellings*.

32. National House-Building Council (1985), Practice Note 3, Building Near Trees.

33. Tomlinson, M. J., Driscoll, R. and Burland, J. B. (1978) *The Structural Engineer*, No. 6, Vol. 56A, pp. 161–73.

34. Nixon, P. J. (1978) *Chemistry and Industry*, 5, 160–4.

35. BRE Digest 250, *Concrete in sulphate-bearing soils and ground waters*.

36. BSI (1973) *Protection of buildings against water from the ground*, CP 102.

37. BSI (1970) *In-situ floor finishes*, CP 204, Part 2.

38. BSI (1972) *Wood flooring (board, strip, block and mosaic)*, CP 201, Part 2.

39. BSI (1984) *Ceramic floor and wall tiles*, BS 6431.

40. BSI (1972) *Tile flooring and slab flooring*, CP 202.

41. BSI (1985) *Code of practice for use of masonry, materials and components, design and workmanship*, BS 5628, Part 3.

42. BSI (1985) *Code of practice for the thermal insulation of cavity walls (with masonry or concrete inner and outer leaves) by filling with urea-formaldehyde (UF) foam systems*, BS 5618.

43. BSI (1985) *Existing traditional cavity construction*, BS 8208, Part 1.

44. BRE Digest 277, *Built-in cavity wall insulation for housing*.
45. Duell, J. and Lawson, F. (1977) *Damp Proof Course Detailing*, The Architectural Press, London.
46. BSI (1976) *Code of Practice for External Rendered Finishes*, BS 5262.
47. BSI (1978) *Specification for metal ties for cavity wall construction*, BS 1243.
48. BRE Digest 228, *Estimation of thermal movement and stresses, Part 2*.
49. BSI (1972) *Precast concrete cladding (non-loadbearing)*, CP 297.
50. Department of the Environment (1973) *Construction No. 6*, Property Services Agency, Croydon, p. 44.
51. Moore, J. F. A. (1984) The Use of Glass-Reinforced Cement in Cladding Panels, *BR 49*, Building Research Establishment, Garston, Watford, Herts.
52. Bonshor, R. B. (1977) Jointing Specification and Achievement: A BRE Survey, Current Paper 28/77, The Station, Garston, Watford, Herts.
53. BSI (1978) *Code of practice for accuracy in building*, BS 5606.
54. BSI (1969) *The structural use of precast concrete*, CP 116.
55. BRE Digest 223, *Wall cladding: designing to minimise defects due to inaccuracies and movement*.
56. BRE Digest 235, *Fixings for non-loadbearing precast concrete cladding panels*.
57. Bonshor, R. B. and Eldridge, L. L. (1974) *Graphical Aids for Tolerances and Fits: Handbook for Manufacturers, Designers and Builders*, HMSO, London.
58. BSI (1978) *Code of practice for wall tiling: external ceramic wall tiling and mosaics*, BS 5385, Part 2.
59. BSI (1982) *Code of practice for glazing for buildings*, BS 6262.
60. Savory, J. G. and Carey, J. K. (1975) *Timber Trades Journal, Wood Treatments Supplement* 295 (5171), 12–13.
61. Department of the Environment (1981) *Flat Roofs Technical Guide*, Property Services Agency, Croydon.
62. BSI (1977) *Specification for roofing felts*, BS 747.
63. Department of the Environment (1979) *Construction No. 31*, Property Services Agency, Croydon, pp. 20–21.
64. BRE Annual Report (1981/2), The Station, Garston, Watford, Herts.
65. BSI (1970) *Roof coverings: mastic asphalt*, CP 144, Part 4.
66. BSI (1979) *Specification for wood chipboard and methods of test for particle board*, BS 5669.
67. BSI (1977) *Specification for expanded polystyrene boards*, BS 3837.
68. BSI (1983) *Coping of precast concrete, cast stone, clayware, slate and natural stone*, BS 5642.
69. Baldwin, R. and Ransom, W. H. (1978) The Integrity of Trussed Rafter Roofs, *BRE Current Paper 83/78*, The Station, Garston, Watford, Herts.
70. Bonshor, R. (1980) *Architects' Journal*, No. 18, Vol. 171, 881–5.
71. BSI (1985) *Code of practice for trussed rafter roofs*, BS 5268, Part 3.
72. BRE Digest 270, *Condensation in insulated domestic roofs*.

73. *Construction* (1981) No. 36, B. & M. Publications, London, pp. 29–30.
74. National Council of Building Materials Producers (1985) *Statistical Bulletin*, May.
75. Freeman, I. L. (1975) *Architects' Journal*, No. 6 Vol. 161, 303–8.
76. *Construction* (1983) No. 42. B. & M. Publications, London, p. 29.
77. Department of Trade (1982) Standards, Quality and International Competitiveness, Command 8621, HMSO, London.

Further reading

Useful information relating to building failures and defects is contained in the following publications.

1. Building Research Establishment Digests.

   Many of these have been collected in four volumes published by HMSO, London. These are:

   (a) *Building Structure and Services*
   (b) *Building Components and Materials*
   (c) *Building Performance*
   (d) *Design and Site Procedures, Defects and Repairs*

2. Defect Action Sheets.

   These are issued by the Housing Defects Prevention Unit, Building Research Station, Garston, Watford, Herts.

3. Feedback Digests.

   These are contained in *Construction*, formerly *DOE Construction*, a quarterly journal of technical information obtainable from the subscriptions department at B. & M. Publications, London.

4. Practice Notes.

   Issued by the National House-Building Council, Chiltern Avenue, Amersham, Bucks.

5. Wood Information Sheets.

   Issued by the Timber Research and Development Association, Hughenden, Valley, Bucks.

Index